MILITARY ROBOTS

Military and Defence Ethics

Series Editors

There is an urgent and growing need for all those involved in matters of national defence – from policy makers to armaments manufacturers to members of the armed forces – to behave, and to be seen to behave, ethically. The ethical dimensions of making decisions and taking action in the defence arena are the subject of intense and ongoing media interest and public scrutiny. It is vital that all those involved be given the benefit of the finest possible advice and support. Such advice is best sought from those who have great practical experience or theoretical wisdom (or both) in their particular field and publication of their work in this series will ensure that it is readily accessible to all who need it.

Also in the series

The Warrior, Military Ethics and Contemporary Warfare
Achilles Goes Asymmetrical
Pauline M. Kaurin
ISBN 978-1-4094-6536-2

When Soldiers Say No
Selective Conscientious Objection in the Modern Military
Edited by Andrea Ellner, Paul Robinson and David Whetham
ISBN 978-1-4724-1214-0

From Northern Ireland to Afghanistan
British Military Intelligence Operations, Ethics and Human Rights
Jon Moran
ISBN 978-1-4094-2897-8

Military Robots

Mapping the Moral Landscape

JAI GALLIOTT
The University of New South Wales, Australia

Routledge
Taylor & Francis Group
LONDON AND NEW YORK

First published 2015 by Ashgate Publishing

Published 2016 by Routledge
2 Park Square, Milton Park, Abingdon, Oxfordshire OX14 4RN
711 Third Avenue, New York, NY 10017, USA

First issued in paperback 2017

Routledge is an imprint of the Taylor & Francis Group, an informa business

British Library Cataloguing in Publication Data
A catalogue record for this book is available from the British Library

The Library of Congress has cataloged the printed edition as follows:
Galliott, Jai.
Military robots : mapping the moral landscape / by Jai Galliott.
pages cm
Includes bibliographical references and index.
ISBN 978-1-4724-2662-8 (hardback) -- ISBN 978-1-4724-2663-5 (ebook pdf) -- ISBN 978-1-4724-2664-2 (epub) 1. Military robots--Moral and ethical aspects. 2. Drone aircraft--Moral and ethical aspects. 3. Military ethics. I. Title.
UG479.G35 2015
172'.42--dc23

2014035748

ISBN 13: 978-1-138-07971-7 (pbk)
ISBN 13: 978-1-4724-2662-8 (hbk)

Contents

Preface

When I resigned my commission from the Royal Australian Navy in 2007, what little public debate existed about the ethics of military robotics was rather inchoate. The sorely underestimated number of civilian deaths and lack of public oversight called for a much more lengthy investigation. Impassioned, I commenced years of intense research. It was toward the end of 2009 that this book started to come together. Fortunately, I was not alone in endeavouring to shed light on this topic. At around the same time, on the other side of the Pacific, a then-graduate student by the name of B. J. Strawser was nearing completion of a PhD on moral responsibility and autonomy in war. Motivated by recent military events, he had decided to add a final chapter to his thesis on what were increasingly prevalent drone wars. Together – along with other similarly concerned scholars such as Armin Krishnan, Christian Enemark, Avery Plaw, Patrick Lin and Robert Sparrow – we helped form the contours of the early academic debate.

This debate has since gained significant momentum. All of a sudden, everyone wants to offer their opinion on the deadly strikes they facilitate, whether it is on political talk shows, in op-eds or congressional hearings. Human rights groups, armchair theorists and newspaper columnists are all warning of the prospect of fully autonomous robots that can conduct lethal military operations without a human in the loop. Some would say that I should be pleased by the invigorated debate, but I am far from so. More often than not, these latecomers have vested interests and reduce the drone debate to questions that demand seemingly impossible answers, 'Do you think drones are good or bad?' or 'Are you for or against the use of drones?' This does little to aid anybody's cause and turns what should be a long academic conversation into a fairly shallow argument that echoes the polarisation found in so much of today's political discourse. The reality is that drone warfare – past, present and future – is more complex than most commentators are typically willing to admit.

While the popular interest has also triggered many in a wider body of academic researchers to jump on the 'drone bandwagon' in the hope of publishing rather than perishing in a tightening higher education environment, this more scholarly debate is scattered across a range of websites, journals and edited books. My aim in this book is to thoughtfully bring together the relevant arguments that have surfaced over the last five years and to carefully examine them under the broad lens of just war theory. Although a work of philosophy is unlikely to have anything other than a minor effect on real-world policy, it should nevertheless be noted that I have written this book with the hope of providing some genuine practical guidance to my former colleagues who lead and protect.

Jai Galliott
Sydney, Australia
April 2014

Acknowledgements

A number of people have helped to make this book possible, but none more so than Mianna Lotz and Aleksandar Pavkovic of Macquarie University, who provided me with the opportunity and much needed encouragement to see this work through to successful completion. Further thanks go to B. J. Strawser, Avery Plaw, Armin Krishnan, Patrick Lin, Rob Sparrow, Katina Michael, Christian Enemark, Peter Singer, Stephen Coleman, John Kleinig, Jeff McMahan, Thomas Hurka, Ed Barrett, Shannon Ford, Seumas Miller, Nicole Vincent, Cynthia Ung, Jeanette Kennett, Alex Leveringhaus, Tjerk de Greef, Heather Roff-Perkins, David Wallace, Anke Snoek, Ruby Catsanos, Thomas Bruneau and Tereza Hendlova, all of whom offered guidance of one sort or another. My gratitude also goes to my mother for providing perspective and, of course, my wife for ensuring that this book did not become all-consuming by distracting me with our engagement, marriage and first child.

I am also grateful to have had the opportunity to test many of my ideas elsewhere. For this honour, I am particularly indebted to: Jeroen van den Hoven for providing funds to attend TU Delft and Oxford University to present what became the basis of Chapter 9; Henrik Syse of the *Journal of Military Ethics* for coordinating feedback on two articles concerning asymmetry that now form part of a heavily revised Chapter 7; Avery Plaw for kindly inviting me into his family home and allowing me to present my application of *jus post bellum* found in Chapter 8 at the University of Massachusetts; and Shannon Ford and Adam Henschke of the Centre for Applied Philosophy and Public Ethics for inviting me to speak on the ethical justifications that later became the focus of Chapter 3. This book is also informed by some of my other related works published in *Dynamique Internationales*, *Ethics*, *The Canberra Times*, *IEEE Technology & Society* and chapters published by Open Court Press and IGI Global.

Finally, I would like to give special thanks to Kirstin Howgate as Publisher for Politics and International Relations at Ashgate Publishing, whose passion for emerging technologies and international affairs prompted me to publish this work. I also appreciate the work of Don Carrick and James Connelly of the University of Hull, George Lucas of the U.S Naval Academy and Paul Robinson of the University of Ottawa, who recognised the value of this project through their very timely and valuable Military and Defence Ethics series.

Chapter 1
Introduction

Many weapons technologies introduced over the past two millennia have been described as making warfare less destructive, horrible and cruel. Generating particular interest and concern of this kind are those that add a 'buffer zone' between friend and foe with the alleged aim of bringing about so-called 'cleaner' deaths and a quicker end to warfare. Consider the fact that our ancestors once relied upon simple rocks and wooden clubs to defeat enemies at very close range. As knowledge was applied to weapons design, spears, bows and arrows, trebuchets and other mechanical artillery were created. This was followed by the development of gunpowder, cannons, firearms and other modern weaponry, all of which further increased the range between the attacker/s and the attacked, supposedly making warfare less horrendous than the close-quarters combat of the past. However, for those who were showered with arrows and bombarded with cannon fire at the Battle of Crécy, or who saw combat in more recent bloody conflicts such as World War II or Vietnam, the notion that the advent of new weaponry could somehow reduce the carnage of war must have been seen as the most sickening and preposterous of suggestions.[1] That is to say that in spite of the positive hype, real questions exist concerning whether emerging military technologies actually deliver a genuine reduction in, or mitigation of, the savagery of war.

These questions are now crucial because the optimism that once surrounded the creation of weapons like artillery – which, in 1621, was mistakenly forecast to bring war to 'quicker ends than heretofore' while avoiding the 'great expense of blood' and reducing the 'number of men slain'[2] – has, to an extent, been renewed with the advent of military robots to be known here as *unmanned systems* (see the next section for a full definition). In other words, developments in unmanned weapons technologies have raised the hope that the inevitable conflicts of the twenty-first century and beyond might be waged in a more humane manner than in the past.[3] The media often conveys this hope in reports detailing the potential of unmanned systems to save the lives of soldiers and civilians. The justifications for the use of these systems will be examined later, but much of this renewed hope and optimism can be traced back to the *first* Gulf War of the early 1990s, which was a technological testing ground for the United States military. With the aid of superior technology, the US achieved a rapid and decisive victory over Saddam Hussein's large army, while only incurring a few hundred American fatalities and minimising other casualties.[4] Years of training and preparation for a technologically advanced war had seemingly paid off quite handsomely and this left the US military with a yearning for more technology, as it thought that it could apply a similar formula to almost any potential scenario for a quick, efficient and decisive victory.[5] As was evidenced by the *second* Gulf War, the formula could not be carried over quite so easily in practice, but the allure was understandable. The task before us is to consider whether it is wise to subscribe to the current optimism that surrounds unmanned systems.

But before going any further, it needs to be said that unmanned systems are not the only modern-day option promising to make warfare more humane. Cyberwarfare – i.e. any sort of state-based action which aims to penetrate another state's computer network/s in order to cause disruption or damage – offers a potentially bloodless alternative to traditional war and has received increased attention in recent years.[6] This is due to Russian attacks on Estonia and Georgia, and probable attacks by China, North Korea and Iran on US military and economic systems.[7] Advocates of nonlethal or less-than-lethal weapons also argue that today's adversaries can be engaged without the need for deadly weaponry. Lasers can dazzle, microwave systems can heat skin, chemicals can incapacitate, kinetic munitions can deliver blunt force and vortex generators can create shockwaves.[8] These technologies lead some theorists to the belief that unmanned systems are a subset of an ongoing 'revolution in military affairs' known as the 'information revolution'. This revolution is driven by the desire for more accurate and timely battlespace information as well as a reduction in bloodshed. This desire has led to many changes in the force structures, strategies and tactics (which are thought to be signifiers of a true revolution) of many military forces.[9] The idea is that with the information and capabilities provided by emerging technologies, military forces can work more effectively and efficiently, minimising the harm caused to all those involved in warfare. There is, however, significant controversy concerning whether there has been, is now, or ever will be, a revolution in military affairs.[10] It could also be argued that we are on the cusp of an even more synergistic revolution that includes unmanned systems, other information communication technologies, biotechnologies and nanotechnology. It is difficult to pinpoint the determining attributes of this purported revolution, but the possibility of such a paradigm shift should not be ignored since, when technologies converge, they have a greater tendency to determine how societies and military forces interact.

While it is not obvious whether unmanned systems are best described as part of the information revolution, some new revolution, or both, it is clear from the adoption rate discussed in the next chapter that military forces seeking to modernise themselves are keen to embrace this technology. History has demonstrated that technology is all too often a double-edged sword.[11] As it comes to provide positive benefits to society, it also acquires a more destabilising potential.[12] Well-known philosophers of technology including Neil Postman,

Carol Mitcham and Jacques Ellul have written about technology's potentially negative influence on humankind and have warned us against blindly embracing any technology. In *Why Things Bite Back: Technology and the Revenge of Unintended Consequences*,[13] Edward Tenner also goes to great lengths to remind us that technologies, which are allowed to mature, tend to replace acute life-threatening problems with slower-acting and more persistent problems. Tenner labels these 'revenge effects' and elaborates that they include rearranging, repeating, recomplicating, regenerating and recongesting effects.[14] These effects do not generate in a vacuum and tend to reach their full potential when technology is impacted by social, cultural, legal and organisational factors,[15] which is why it is important to think of unmanned systems in the wider developmental picture. But to be clear, the risk of unintended consequences does not mean that we should cease weapons development altogether. Many technologies have produced undesirable consequences, but we have not stopped using them. In truth, it is a matter of long held contention as to whether weapons development has overall made war more or less destructive. Therefore, without comprehensive research and analysis, it would be a mistake to conclude that there has been no improvement.

The question to ask is whether it is possible to manage and transform the techno-social and ethical challenges associated with existing weapons into practical and beneficial solutions. This question must also be extended to unmanned systems because, as with other technologies, they are not of unqualified virtue. Incidentally, this message is ingrained in the origin of the term 'robot'. After World War I, there was renewed interest in using machines to perform labour on behalf of humans, given that a whole generation of industrial workers had been incapacitated by weapons or struck down by influenza. Karel Capek first used the term 'robot' – derived from the Czech word 'robota' for 'labour' – in his 1920s play *Rossum's Universal Robots* (or *R.U.R.*).[16] Capek's play envisioned service robots becoming ubiquitous in society. Unsurprisingly, the robot servants eventually become dissatisfied with their role, revolting against their human masters. This vision gave rise to a whole new genre of science fiction works expressing negative sentiments about this dystopian future.[17] Popular culture films such as *The Day the Earth Stood Still*, *Eagle Eye*, *Star Wars*, *Terminator* and *Transformers*, to name only a few, also tell cautionary tales about robots in the line of defence. More often than not, they convey concerns about errors, emergent behaviour and insufficient programming.[18] However, for some, these popular warnings have had the unfortunate effect of normalising the notion of robots capable of independent action and potentially trivialising some of the relevant concerns. This book seeks to counter this effect and furthers the cautionary tale in a more scholarly tone, style and format. It examines the reasons for using unmanned systems and will look at the constraints that rightly apply to their use, as well as matters pertaining to their social, strategic and moral impact. The overarching objective is to draw attention to the human role in unmanned warfare and demonstrate that while there is an obligation to utilise unmanned systems, their deployment should be subject to strict oversight and international regulation.

Terms and Definitions

An accurate and judicious assessment of unmanned systems requires that we first have to hand a clear definition and understanding of that technology. The popular discourse on robots has resulted in confusion concerning the different features and classes of robots, and the *Springer Handbook of Robotics*[19] acknowledges that the terms surrounding military robotics are often misleading and poorly defined. Apart from the popular cultural expectations, this has much to do with the many unfortunate, but somewhat necessary, 'divides' and 'disconnects' between the humanities and the so-called 'hard sciences'.[20] In the latter, there is a tendency (or need) to provide focused assessments and definitions that contrast with the broader analyses and definitions often given in the 'softer' humanities. It also has something to do with the fact that many of the World's major military forces have now commissioned reports into the state of the various technologies involved, and each of these have their own list of terms, definitions, classifications and other military jargon, which then make their way into the academic debate. This makes any sort of discussion about military robotics technologies slightly more difficult in terms of locating the appropriate literature, and more confusing in the sense that one then has to compare the different terms and their definitions. As with any emerging topic, the definitions and ethical issues have evolved with the increased attention. It is impossible to standardise all the terms and problems; instead, the aim here is to point out some of the possibilities and outline the working definitions that will be adopted throughout this book.

To begin with, we must ask what seems to be a fairly basic question: what is a robot? As mentioned above, we have long read and seen cautionary tales concerning robotics in society and the military. As such, it might be thought that asking such a question is redundant as the answer is reasonably obvious. However, the reality is quite to the contrary. To this day, there is still no consensus concerning the definition of a robot, even amongst roboticists themselves (let alone between roboticists and philosophers). Some people consider a robot to be a mere machine with sensors and actuators that allow it to interact with the external world.[21] However, any computer with a printer or a DVD drive would qualify under this definition, and this is hardly what most people have in mind when they think of a robot.[22] From an engineering point of view, a robot involves the embodiment of manipulative, locomotive, perceptive, communicative and/or cognitive abilities in an artificial body, which may or may not have a humanlike appearance.[23] The more philosophical approaches tend to define a robot as a machine that 'senses', 'thinks' (or 'decides') and 'acts'.[24] Of course, the question of whether the abilities to sense, think, decide or act can accurately be ascribed to a machine is a matter of significant contention. This matter aside, a generally agreed upon requirement is that a robot must have a degree of technical autonomy which, very simply described, is an ability to operate unsupervised. This necessitates that there be a way of receiving and processing data from the operating environment and an ability to interact with its surroundings.

Many of today's unmanned systems fulfil these criteria, but there are some that are only outfitted with sensory gear and processors to enable a distant human to operate the device. The consequence of this is that, unless the definitional reach is extended to include the human operator's decision-making capabilities, it cannot be said that all unmanned systems are robots in the strict sense of the term. It is partly because of this ambiguity concerning the common elements that a distinct definition of 'unmanned systems' is necessary.[25] For the sake of this book, 'unmanned systems' will be used interchangeably with the term 'drones' and defined relatively broadly and uncontroversially as:

> **Defn:** a group of powered electro-mechanical systems, all of which have in common that they:
>
> 1) do not have an onboard human operator;
>
> 2) are designed to be recoverable (even though they may not be used in a way that renders them such); and,
>
> 3) in a military context, are able to exert their power in order to deliver a lethal or nonlethal payload or otherwise perform a function in support of a military force's objectives.[26]
>
> They include categories of unmanned ground vehicles, unmanned maritime vessels (surface and underwater), unmanned aerial vehicles and unmanned space vehicles (although the latter category may be considered a subset of the other three).[27]

It must be acknowledged that the term 'unmanned' is actually a misnomer in the sense that this type of warfare remains a manpower intensive endeavour despite the absence of an onboard operator, due primarily to the number of support personnel involved.[28] Also note that the word 'systems' does not just refer to the different vehicles and vessels; it also includes the control systems that allow the various platforms to operate, such as satellites and ground stations. Some would argue that weapons like missiles, rockets and unattended munitions (e.g. landmines) *are* unmanned systems.[29] However, for the purpose of this book, these weapons will *not* be considered as such. Some versions of these weapons may be a constituent part of an unmanned system's weapons arsenal (for example, a missile or rocket on an unmanned aerial vehicle); therefore, it does not make great sense to think of them as unmanned systems in themselves. Furthermore, they are not recoverable: they are destroyed on impact or detonation. Finally, many of these weapons – some missiles and landmines, for instance – are not actively powered for all stages of their operation. All of these characteristics are particularly important to military forces, which need clear classifications in order to adhere to (or perhaps circumvent) weapons control regimes.

Beyond these largely technical points, there are also more conceptual reasons for limiting the application of the definition. As mentioned above, the motivation behind the development and use of unmanned systems has much to do with distancing the warfighter from the conflict zone. In the early life of these systems, this typically meant replacing a manned system with the unmanned equivalent, such as replacing a fighter aircraft with an unmanned aerial vehicle. On the other hand, rockets, missiles and unattended munitions do not have clear unmanned equivalents. Although they perform tasks that a human might otherwise have to perform via other means, the link to current unmanned systems is much more tenuous. Admittedly, not all of today's unmanned systems are designed for uses analogous to that of a manned equivalent, but this is the way they were originally conceived and this conception still proves useful in the great majority of modern cases. Additionally, weapons such as landmines and nuclear missiles are generally indiscriminate by design, often causing pain and suffering long after the war in which they were used has ceased. This makes them particularly abhorrent and worthy of independent moral consideration. It is also questionable whether any of these weapons should be considered to be robots.[30] For all of these practical and conceptual reasons, these weapons will not be considered unmanned systems. At best, some complicated versions of these weapons may be considered to be 'quasi-unmanned systems' (defined here as those systems that were influential in the design of true unmanned systems, but do not fit with the definition above or otherwise deserve separate treatment for one reason or another). Of course, this is not to deny that these weapons (along with other weapons that facilitate killing at a distance) raise ethical issues that are exacerbated and/or complicated by unmanned systems proper.

What remains is a need for a more comprehensive outline of the modes of operating unmanned systems. Let us temporarily set aside the many philosophical complexities surrounding the term 'autonomy' and its links to debates about agency, the mind, cognition and so on. For the moment, all that is needed is a technical definition that accounts for the removal of the human being from the control loop and increasing levels of unsupervised operation among unmanned systems. As autonomy is a matter of degree and can range from being low (namely where the supervisor is very much 'in the loop') to high (where the supervisor is 'on the loop' or virtually totally removed from the loop), it makes sense to think of a continuum of autonomy, along which the various platforms fall, depending on their specific design and capabilities.[31] However, for the purpose of examining the associated ethical issues, it helps to draw some lines along the continuum to create clearly demarcated levels of autonomy, even if there is no need to explicitly call on these levels. Of course, one question concerns how many lines should be drawn. While some divide the continuum into 15 to 20 different levels of autonomy, four degrees of separation are provided here, affording a clearer and more focused indication of what is most fundamental about the different types:

Autonomy Level 1 – Non-autonomous/Teleoperated:
A human operator controls each and every powered movement of the unmanned platform. Without the operator, teleoperated systems are incapable of effective operation.

Autonomy Level 2 – Supervisory Autonomy:
A human operator specifies movements, positions or basic actions and the system then goes about performing these. The operator must provide the system with frequent input and diligent supervision in order to ensure correct operation.

Autonomy Level 3 – Task Autonomy:
A human operator specifies a general task and the platform processes a course of action and carries it out under its own supervision. The operator typically has the means to oversee the system, but this is not necessary for extended operation.

Autonomy Level 4 – Full Autonomy:
A system with full autonomy would create and complete its own tasks without the need for any human input, with the exception of the decision to build such a system. The human is so far removed from the loop that the level of direct influence is negligible. These systems might display capacities that imitate or replicate the moral capacities of sentient human beings (though no stand on this matter shall be taken here).

While the second and third levels of autonomy could possibly be merged together as they are both levels of semi-autonomous operation, the distinction is nevertheless retained since quite a significant shift can be noted in the role of the human operator between the two levels.

The Undeveloped Arguments

As noted, the literature on unmanned systems has grown substantially over the past few years as the use of these systems has expanded, but much of it is of a strictly technical or legal nature. In fact, one of the most surprising features of the military robotics debate is that there are few scholarly treatments that deal with the associated ethical issues in the detail that they deserve, particularly when you compare this technology to those in other fields such as medicine, where they have received a great deal of attention and attracted much research funding. In what follows, the state of the military 'roboethics'[32] literature is briefly surveyed in the hope that it will highlight the need for more comprehensive research bringing together the different issues and examining the links between them.

Perhaps the most popular book to broach the ethical issues associated with unmanned systems is Peter W. Singer's *Wired for War: The Robotics Revolution and Conflict in the 21st Century*.[33] Singer, who should not be confused with the famous Australian utilitarian, is probably best described as a political theorist and has previously written about the problems associated with the military exploitation of children and the rise of private military contractors. *Wired for War* aims to demonstrate that unmanned systems have fundamentally changed the way war is carried out and touches on a number of very important ethical concerns relating to psychological issues, terrorism and the nature of the warrior and warfare, among others. However, it is very much part of the 'first wave' of the drone debate, characterised by journalistic flair, science fiction references and the presentation of problems in 'laundry list' fashion.[34] His aim was to provide an introduction to issues associated with the rise in robotics technology and war, rather than to substantively engage with the more thorny problems.[35] Singer's main contribution comes in the extensive primary research he conducted, interviewing serving soldiers, statesman and weapons designers. This book draws on this wealth of information in tackling some of the relevant problems on a deeper level.

Spurred on by Singer's work and the enhanced public profile of the debate, a small but dedicated group of concerned ethicists has also been working on the ethical issues associated with unmanned systems, many starting well before the term 'drone' had entered the public consciousness. Between them, they have published an assortment of journal and magazine articles, book chapters and a lengthier military report.[36] They have also presented numerous papers at events such as the annual International Society for Military Ethics conference and workshops hosted by the International Committee for Robot Arms Control, among others. Armin Krishnan's *Killer Robots: Legality and Ethicality of Autonomous Weapons*[37] forms the latter part of this first wave and despite being the broadest existing work on military robotics, suffers from limitations in the scope of its coverage. For instance, while warfare is typically a deadly enterprise, not all unmanned systems are used in a lethal capacity as implied by the term 'killer robots', nor are they all autonomous. Furthermore, the discussion of ethics is relegated to one chapter and arguments concerning responsibility are not sufficiently addressed. What is most concerning, however, is that the focus of this early scholarship has shaped the debate in an unusual way. The narrow concentration on the potentially dire consequences of autonomous systems, coupled with a lack of detailed philosophical examination offered by some of the relevant resources, has had the unfortunate effect of polarising writers and readers into different camps for or against these systems. That is, the first wave has inadvertently turned the academic conversation into a fairly shallow argument echoing much of today's political discourse.

Some participants have recognised the importance of moving beyond the first wave of the drone debate and pushing the conversation to a more analytic level. Christian Enemark's *Armed Drones and the Ethics of War: Military Virtue in a Post-Heroic Age*[38] is the first book of the second wave. This important work on drones and the requirement of mutual risk for combatants draws on and closely mirrors the argument put forward in Chapter 7. But while many of Enemark's worries are shared here, he does not get to the bottom of the justification for using these systems or deal with matters of responsibility, amongst others. Bradley J. Strawser's *Killing by Remote Control: The Ethics of an Unmanned Military*[39] is also part of the deeper analytic push and purports to improve the analysis in concrete and specific ways, namely by differentiating between the *practical* and the *in principle* concerns. Of course, many of the individual contributions invite detailed responses, particularly Strawser's chapter on the permissibility of unmanned warfare and Robert Sparrow's on the moral implications of significant imbalances in technological strength. This book provides these responses along with a clear, comprehensive and unifying academic analysis of all the major ethical aspects surrounding unmanned systems. It carefully examines the moral imperative behind the use of these systems, the moral constraints that rightly apply and, of course, the ethical concerns that are exacerbated or generated by their use. This involves synthesising the existing fragmented literature, subjecting it to critical thought, expanding upon the areas that have been neglected and establishing the links between the different justifications and moral concerns. In assessing the permissibility of drone deployment,[40] this book also hopes to provide a degree of normative guidance. Given that we live in a risk-focused society and cannot uninvent these systems, it might also be labelled a work of 'consequence management' and will be grounded mainly, although not exclusively, in the Anglo-American philosophical tradition.[41]

On a final methodological note, it should be said that it is very difficult to divide theory and practice as far as unmanned systems are concerned. Much of the drone debate is about determining the morally acceptable level of risk in any given scenario. When we ask whether a commander should authorise an unmanned bombing that involves civilian losses, or whether it is justifiable to fight a war with virtual impunity, we are asking distinctly moral questions to which it is difficult to provide definitive answers. However, many are also linked to empirical facts of ethical importance: are these systems more precise than their manned equivalents? Do they limit the psychological harm caused to soldiers? Can these systems be programmed to perform better than humans? Of course, as unmanned systems are still in their operational infancy when considered in the grand scheme of military-technical advancement, there is little hard data with which to inform judgement. For this reason, no apology is made for mixing philosophical theory with the best available evidence from experienced military practitioner-scholars, albeit occasionally anecdotal.

The Way Forward

Having only briefly outlined the potential of unmanned systems in terms of reducing the savagery of war and introducing a variety of unintended consequences, the first few chapters of this book examine precisely what unmanned systems are in some technical detail, as well as why states and their military forces feel compelled to use them and how they ought to go about this in theory and practice. More specifically, Chapter 2 provides a technological survey of unmanned systems, spanning from their ancient origins to future developments, with a focus on the progression of autonomy-capacity over this period. Chapter 3 examines the operational justifications given for the use of these systems and establishes the constant search for efficiency and effectiveness as a moral imperative for any legitimate military force through an application of social contract theory. With the case for unmanned systems use put forward and critiqued, Chapter 4 establishes the need for restraint in unmanned warfare and outlines a contractually grounded just war theory as the practical moral framework of choice.

The remaining chapters of the book examine the actual and potential challenges associated with the employment of unmanned systems technologies and unmanned warfare writ large. From both within and beyond the just war framework of the previous chapter, they consider the efficacy of unmanned systems in combat and question whether these technologies are indeed the more humane weapons or 'silver bullets' that are touted to be key to the effective and efficient application of military force. To this end, Chapter 5 focuses on the relevant technological and operational dimensions of unmanned warfare. It is asked: do unmanned systems keep people out of harm's way? Are their operators prone to combat complacency? What are the technical limitations? It draws these questions together in the context of risk transference and considers to what extent this is ethically permissible. This discussion is continued in Chapter 6, but with a focus on the more psychological, social and political dimensions of unmanned warfare and their impact on the principal war-making agents, namely: the public, unmanned systems operators, their commanders and higher-level military and political decision makers. The chapter concludes by considering whether more advanced technology represents the solution to some of the identified challenges.

The potential for a technological solution is further problematised in Chapter 7, which examines the concept of asymmetry as it pertains to unmanned warfare. It is argued that drone warfare is so radically asymmetric in some instances and in comparison to other forms of warfare, that it sets justice and fairness in conflict or competition with the moral and strategic aims of drone wars. Chapter 8 looks at the effect of drone wars on local populations and considers how a greater concern for the ethical considerations relevant to the post-conflict environment may, in the case of the recent conflicts in the Middle East, have dramatically altered the way events unfolded. This is illustrated as an example for future operations and a number of more constructive unmanned systems measures are covered. The final and perhaps most important chapter deals with the attribution of responsibility for the problems and challenges detailed throughout. It analyses a number of barriers to our traditional framework of responsibility and lays the groundwork for a more practically focused and forward looking account aimed at ensuring that unmanned systems receive the attention they deserve from all stakeholders.

Notes

1 Charles Dunlap, *Technology and the 21st Century Battlefield: Recomplicating Moral Life for the Statesman and the Soldier* (Carlisle: Institute for Strategic Studies, 1999), 1.

2 Henry Alford, *The Works of John Donne, with a Memoir of His Life* (London: John W. Parker, 1839), 58.

3 There is significant controversy concerning whether the attacks that characterise twenty-first century warfare are best classified as an element of 'regular warfare' or 'terrorism'. For the sake of this work, it need only be noted that assassination can be a legitimate part of regular military operations and need not be considered extrajudicial killing.

4 James S. Corum, 'Development of Modern Counterinsurgency Theory and Doctrine', in *The Ashgate Research Companion to Modern Warfare*, ed. George Kassimeris and George Buckley (Farnham: Ashgate, 2010), 46.

5 Liang Qiao and Xiangsui Wang, *Unrestricted Warfare: China's Master Plan to Destroy American* (West Palm Beach: NewsMax Media, 2002), 77; Simon Cooper, 'High-Tech War and the Securing of Consent', *Southern Review: Communication, Politics & Culture* 37, no. 3 (2005): 76.

6 Myriam Dunn Cavelty, 'Cyberwar', in *The Ashgate Research Companion to Modern Warfare*, ed. George Kassimeris and George Buckley (Farnham: Ashgate, 2010), 129.

7 Randall R. Dipert, 'The Ethics of Cyberwarfare', *Journal of Military Ethics* 9, no. 4 (2010): 384.

8 Christopher Coker, *Ethics and War in the 21st Century* (New York: Routledge, 2008), 130.

9 Elinor C. Sloan, *The Revolution in Military Affairs: Implications for Canada and NATO* (McGill-Queen's University Press, 2002), 3–4.

10 For varying views on this debate, see Max Boot, *War Made New: Technology, Warfare, and the Course of History, 1500 to Today* (New York: Gotham Books, 2006); Jeremy Black, 'The Revolution in Military Affairs: The Historian's Perspective', *Journal of Military and Strategic Studies* 9, no. 2 (2006/7), www.jmss.org/jmss/index.php/jmss/article/download/111/122; M. G. Vickers and R. C. Martinage, *The Revolution in War* (Washington, DC: Center for Strategic and Budgetary Assessments, 2004); Bernard Loo, 'Revolutions in Military Affairs: Theory and Applicability to Small Armed Forces', in *Military Transformation and Strategy: Revolutions in Military Affairs and Small States*, ed. Bernard Loo (London: Taylor & Francis, 2008); Williamson Murray, 'Thinking About Revolutions in Military Affairs', *Joint Force Quarterly* 16 (1997).

11 Patrick Lin, George Bekey, and Keith Abney, *Autonomous Military Robotics: Risk, Ethics, and Design* (San Luis Obispo: California Polytechnic State University, 2008), 2.

12 Kip P. Nygren, 'Emerging Technologies and Exponential Change: Implications for Army Transformation', *Parameters: The US Army War College Quarterly Senior Professional Journal* 32, no. 2 (2002): 89.

13 Edward Tenner, *Why Things Bite Back: Technology and the Revenge of Unintended Consequences* (New York: Vintage Books, 1996).

14 Ibid.

15 Ibid., 7.

16 Karel Capek, *R.U.R (Rossum's Universal Robots)*, trans. D. Wyllie (Fairford: Echo Library, 2010).

17 Paul Springer, *Military Robots and Drones* (Santa Barbara: ABC-CLIO, 2013), 10.

18 Patrick Lin, Keith Abney, and George Bekey, 'Robot Ethics: Mapping the Issues for a Mechanized World', *Artificial Intelligence* 175, nos 5–6 (2011): 943.

19 Bruno Siciliano and Oussama Khatib, eds, *The Springer Handbook of Robotics* (Dordrecht: Springer, 2008).

20 Peter W. Singer, 'The Ethics of Killer Applications: Why Is It So Hard to Talk About Morality When It Comes to New Military Technology?', *Journal of Military Ethics* 9, no. 4 (2010): 301.

21 Gianmarco Veruggio and Fiorella Operto, 'Roboethics: Social and Ethical Implications of Robotics', in *Springer Handbook of Robotics*, ed. Bruno Siciliano and Oussama Khatib (Dordrecht: Springer Verlag, 2008), 1502.

22 Lin, Abney and Bekey, 'Robot Ethics: Mapping the Issues for a Mechanized World', 943.

23 Ming Xie, *The Fundamentals of Robotics: Linking Perception to Action* (River Edge: World Scientific, 2003), 8.

24 Peter M. Asaro, 'How Just Could a Robot War Be?', in *Current Issues in Computing and Philosophy*, ed. Philip Brey, Adam Briggle, and Katinka Waelbers (Amsterdam: IOS Press, 2008), 51.

25 For a few definitions, see Douglas W. Gage, 'UGV History 101: A Brief History of Unmanned Ground Vehicle (UGV) Development Efforts', *Unmanned Systems Magazine* 13, no. 3 (1995); Laszlo Vanya, 'Excepts from the History of Unmanned Ground Vehicles', *AARMS* 2, no. 2 (2003); United Kingdom Ministry of Defence, 'Joint Doctrine Note 2/11: The UK Approach to Unmanned Aircraft Systems' (Wiltshire: Ministry of Defence, 2011).

26 While there are many 'dual use' problems associated with unmanned systems, only those with strong implication for the military domain are considered here.

27 Each of these categories includes various individual platforms, examples of which are discussed in Chapter 2.

28 Nancy J. Cooke and Roger A. Chadwick, 'Lessons Learned from Human-Robotic Interactions on the Ground and in the Air', in *Human-Robot Interactions in Future Military Operations*, ed. Michael Barnes and Florian Jentsch (Farnham: Ashgate, 2011), 356.

29 Noel Sharkey, 'Moral and Legal Aspects of Military Robots', in *Ethical and Legal Aspects of Unmanned Systems: Interviews*, ed. Gerhard Dabringer (Vienna: Institute for Religion and Peace, 2010), 47–48.

30 For an example of argument that advocates considering improvised explosive devices as robots, see Patrick Hew, 'The Blind Spot in Robot-Enabled Warfare Deeper Implications of the Ied Challenge', *Australian Army Journal* 7, no. 2 (2010).

31 Asaro, 'How Just Could a Robot War Be?', 52; Patrick Hew, 'C2 Design for Ethical Agency over Killing in War', in *The Evolution of C2: Where Have We Been? Where Are We Going?* (Santa Monica: International Command and Control Research and Technology Symposia, 2010), 4; Armin Krishnan, 'Automating War: The Need for Regulation', *Contemporary Security Policy* 30, no. 1 (2009): 176.

32 John P. Sullins, 'Introduction: Open Questions in Roboethics', *Journal of Philosophy and Technology* 24, no. 3 (2011): 233.

33 Peter W. Singer, *Wired for War: The Robotics Revolution and Conflict in the 21st Century* (New York: The Penguin Press, 2009).

34 Bradley Jay Strawser, 'Introduction: The Moral Landscape of Unmanned Weapons', in *Killing by Remote Control: The Ethics of an Unmanned Military*, ed. Bradley Jay Strawser (New York: Oxford University Press, 2013), 4.

35 George Lucas, 'Postmodern War', *Journal of Military Ethics* 9, no. 4 (2010): 292.

36 Lin, Bekey and Abney, 'Autonomous Military Robotics: Risk, Ethics, and Design' (San Luis Obispo: California Polytechnic State University, 2008).

37 Armin Krishnan, *Killer Robots: Legality and Ethicality of Autonomous Weapons* (Farnham: Ashgate, 2009).

38 Christian Enemark, *Armed Drones and the Ethics of War: Military Virtue in a Post-Heroic Age* (New York: Routledge, 2013).

39 Bradley Jay Strawser, ed. *Killing by Remote Control: The Ethics of an Unmanned Military* (New York: Oxford University Press, 2013).

40 While allusions may be made to the moral permissibility of participating in the development of these systems, this is not the primary concern of this book. However, these systems do present a range of interesting ethical questions for engineers and other stakeholders.

41 Coker, *Ethics and War in the 21st Century*, xiii.

Chapter 2
The Rise of Unmanned Systems

For as long as humans have fought wars, they have sought more effective and efficient advantage conferring-technologies.[1] Military forces and their research agencies have therefore dedicated significant resources to the task and the result can be seen in history as a long chain of advances in military technology. Unmanned systems are at the end of this chain and perform target, decoy, logistics, rescue, explosive ordinance disposal, reconnaissance, and combat roles across land, sea, air and space.[2] For many nations, the idea of substituting a machine for a human warfighter in these roles on the modern battlefield is simply too alluring. Aside from preserving the lives of its warfighters and reducing other costs, as will be discussed in Chapter 3, any state that adopts unmanned technologies may also stand to obtain a tactical advantage. The US currently leads in the design, development and deployment of unmanned systems, but 55 other nations are proceeding in similar fashion and this number is steadily increasing.[3] Those using or planning to use drones include Argentina, Australia, Austria, Belgium, Brazil, Bulgaria, Canada, Chile, China, the Czech Republic, Denmark, Finland, France, Germany, Greece, Hungary, Italy, India, Indonesia, Israel, Iran, Japan, Jordan, Lebanon, Latvia, Malaysia, Mexico, Morocco, the Netherlands, New Zealand, Norway, Pakistan, Peru, the Philippines, Poland, Portugal, Romania, Serbia, Singapore, South Africa, South Korea, Russia, Spain, Switzerland, Sweden, Taiwan, Turkey, Thailand, Ukraine, the United Arab Emirates, the United Kingdom, Uruguay and Vietnam.[4] Therefore, it must not be assumed that the American perspective on these systems and the related questions is the only one that is relevant. However, as the US is the largest user of unmanned platforms and arguably the country most prone to using military force beyond its own borders, some special attention must be paid to its technology and the way it is embraced.

In recent years, the US Congress has treated the development and employment of unmanned systems as an area of priority, with several congressional directions on the topic being passed into public law. In 2000, Congress set two major goals in an attempt to integrate unmanned aerial and ground vehicles into the military force structure. The first goal mandated that by 2010, one third of operational deep strike aircraft be unmanned, and the second mandated that by 2015, one third of the Army's operational ground combat vehicles also be unmanned.[5] In 2006, Congress then called for the Department of Defense to establish a policy that would establish a preference for unmanned systems in new weapons acquisitions and address the need for the joint development of these systems and their associated components.[6] While some of these congressional mandates have since been relaxed (not only because the US needs to make substantial cuts to its military spending in order to avoid reaching its looming debt ceiling, but also because of the age-old tendency to cut military research and development in times of austerity) the on-going importance of unmanned systems is reflected in the budget of the Department of Defense. Indeed, there is a strong funding allocation for the acquisition and maintenance of such systems.[7] Take unmanned aircraft as an example. In the 2014 financial year, the US military set aside a total of 3.7 billion dollars for acquiring new drones, excluding ongoing research and development costs.[8] This figure is reminiscent of past financial years, with 3.5 billion requested in 2013 and 4.5 billion in 2012.[9] While only representing roughly 1 per cent of the defence budget, the expenditure on these systems is significant. To put this into perspective, this one per cent is close to the entire annual budget of the US Environmental Protection Agency, which is charged with protecting the health of the physical environment of over 300 million people.[10]

Given the interest in unmanned systems from nations around the globe and the predictable way in which military forces tend to embrace technology, it seems reasonable to expect that these systems will have a notable presence in military force structures for the foreseeable future. However, without a solid understanding of the systems involved, one cannot really begin a detailed investigation of the ethical issues involved. This chapter is therefore dedicated to looking at the rise of unmanned systems and their predecessor technologies. While not exhaustive, the history given here will suffice to convey a good sense of the general trajectory of these systems, which reflects their becoming increasingly autonomous and field capable. The development of these systems will be considered in four key stages. The first examines the history of unmanned systems in early thought and mythology. The second section looks at the history of more modern unmanned systems and associated technologies. The third outlines the systems in use today across the various military domains – firstly land, then sea and finally air. The fourth anticipates the future direction of unmanned systems research.

Origins in Early Thought and Mythology

As already noted, the rapid rise of unmanned systems has taken many by surprise. This level of surprise can only be attributed to the fact that the evolutionary history of this revolutionary technology tends to be ignored. Unmanned systems are often described as being a product of modern times, but this is not entirely true. Nearly every technology results from a long trail of small technological developments and unmanned systems are no exception.[11] But the technological developments considered in the following sections are also the result of something else, namely inventiveness. Since inventiveness is always preceded by the power of creative thought and imagination that is (for now) unique to humankind, it is also important to look at the early thought surrounding robotic systems and distance weaponry. It might be thought that this is not such an important task, but the genesis of unmanned systems goes back many millennia. The only way to begin to examine such a long history is through recounting ancient myths and stories.[12] Even if the chains of history are somewhat broken, it is worthwhile recounting them, as they have given today's engineers the motivation to seek to match those technological and engineering feats spoken of by the storytellers. It must be remembered that many weapons, including the atomic bomb, first had their place in story.

While the term 'robot' was coined in 1920, the quest to automate things dates back much further.[13] Homer's *Iliad* recounts myths surrounding the life of Hephaestus, which incorporate (or perhaps conjure up) the idea of complex intelligent machines and automata (mechanical humanoid creatures), and was written some time between 1200 and 800 BC and drew on legends dating from as far back as 3500 BC.[14] Hephaestus, purportedly the son of Hera and Zeus, was the god of technology and blacksmiths. He was born with both a lame leg and a clubfoot, and because of this, Hera eventually banished him to a volcano on the island of Lemnos. There he built a palace and forged all sorts of metal creations.[15] Many of these were designed to help him with his handicap. There are stories about tripod-type devices that were capable of wheeling themselves around a room under their own power. Originally intended for food delivery at banquets, these tripods would have presumably been very useful for carrying Hephaestus's tools, wheeling people around and so on.[16] Another of Homer's stories suggests that Hephaestus had created animate helpers that acted:

> … in support of their master … These are golden, and in appearance like living young women. There is intelligence in their hearts, and there is speech in them and strength, and from the immortal gods they have learned how to do things. These stirred nimbly in support of their master …[17]

Hephaestus is said to have forged these handmaidens from gold to serve as workshop servants. These stories and the associated machine conceptions could be considered the ancient equivalent of modern day care robots.[18]

Not all of Hephaestus's supposed creations were conceived for his own personal care, however. He was also a blacksmith to the gods, creating their weapons and armour. One of his bittersweet creations was Talon.[19] As the myth goes, Zeus asked Hephaestus to create an automaton for one of his sons, Minos, the King of Crete. Hephaestus fulfilled Zeus's request and built Minos a giant animate bronze man, Talon. Talon is occasionally depicted as being winged and capable of flight,[20] and while not confirmed by mythological reference, this would make sense as Talon is said to have made his way around the island three times a day protecting Minoan Crete from would-be intruders, throwing stones at anyone that presented a threat.[21] A number of other mythological creations are also attributed to Hephaestus. He made a flying boat out of gold for the sun god Helios, which was capable of taking him from the West (the place of his setting) to the East (the place of his rising), all 'as he sleeps soundly'.[22] He is also purported to have created a metal eagle which, at Zeus's command, feasted daily on the ever-regenerating liver of the chained-down Prometheus as punishment for stealing food and fire from the gods and giving it to mankind.[23] His other creations include animal-like automatons, apparently made for King Aeetes, son of Helios. Among these was a pair of watchdogs, one of gold and one of silver, which were guardians of the King's palace. These automaton dogs were said to be 'deathless for ever and unageing'.[24] Similarly, there were bronze bulls capable of breathing out blazing flame, which Jason and the Argonauts had to yoke before King Aeetes would hand over the legendary Golden Fleece.[25]

Stories of mythological devices and distance weaponry are not just found in the ancient Greek literature. There is evidence that similar concepts had also invaded early Asian thought. An ancient Chinese myth tells us that while on a tour of inspection, King Mu of Chou (~1023–957 BC) had an artificer by the name of Yen Shih presented to him. Yen later presented a mechanical man to the king, one constructed of mere leather, wood, glue and lacquer, yet holding a variety of artificial organs. The automaton is recorded to have amazed the king by taking great strides and speaking the human word.[26] The Hindu tradition also records elements of the technological dream that has become a twentieth century reality. In various places in the *Mahabharata*, one of two major Sanskrit epics of ancient India dating from around 400 BC, it is stated that Krishna's enemies sought the assistance of certain demons to build an aerial chariot with sides of iron and clad with wings.[27] The chariot was flown through the sky until it reached the area where Krishna's followers dwelt and, from there, missiles were hurled down from the chariot, destroying everything in their path.[28]

Beyond the very early mythological references, it is worth noting that there exist many other records which describe technological innovations that are much more realistic in nature (although almost certainly influenced by many of the ancient texts cited above). It is fairly commonly known that the Chinese were amongst the first people to utilise the kite, but lesser known is that kites were occasionally used as weapons in war. While obviously not mechanical, kites clearly add a distance between warring parties. In 200 BC, a Chinese general by the name of Han Hsin is said to have flown a kite over the wall of his enemy's fortress, using the string to measure his distance from the enemy.[29] This intelligence would then have been used to plan and coordinate a ground attack. Some say that kites were used as an early form of aerial bombardment: incendiary devices were attached to the kites, which were flown over the enemy's walls to destroy wooden buildings.[30]

Furthermore, there are some surviving construction plans documenting early machines that are thought to have existed in the real world. Among them was a wheeled cart-like device made by Hero of Alexandria in approximately 60 AD.[31] It was driven forward by an ingeniously simple falling weight that was attached to a rope wrapped around its axle, turning the wheels as the weight fell with gravity. If the rope was wrapped in the opposite direction, it could also produce a backward motion. If a segment of rope were left hanging, this would produce a pause in the device's operation and if it had independent wheels there could have been more options. This was perhaps the first proto-robot and remarkable in that it was in some sense programmable.[32] Leonardo da Vinci took this design even further, designing a similar robot in the form of a knight. From da Vinci's recently discovered blueprints and subsequent attempts to produce what was sketched, it is evident that the knight would have possessed the ability to sit up, wave its arms, grasp large objects, move its head and open its mouth.[33] Much like Hero's rope-driven system, Leonardo's *Knight* operated via a more complex system of cables and while the elementary programming of this system would have been difficult, there was obviously much potential. Da Vinci also raised the spectre of flying machines, with a single helical (or spiral-like) rotor-powered device designed to compress air and produce lift.[34]

Other precursor technologies are simply too numerous to mention. This is partly due to the fact that the wealthiest members of ancient society came to see these contraptions as an interesting form of entertainment and engaged the services of other inventors, engineers and mathematicians to produce many examples and what could be considered a proto-robotics industry. Without detailing all of these, it is important to acknowledge the work of the visionaries of our distant past who produced concepts and inventions that came to play a central part in the earliest stages of the development of unmanned systems. Some understood that with the right technology, unmanned systems amount to more than just entertainment and that these devices could be harnessed to confer an advantage in conflict. While these inventors lacked the technology to develop their creations further, they nevertheless made a conceptual contribution that heavily influenced those inventors and technologies that were to follow.

Modern Developments: From Teleoperation to Supervisory Autonomy

Moving forward, World War I is widely taken to be the starting point for discussions of the modern history of unmanned systems, but important influences date back to as early as 1860 and the American Civil War. In this war of independence, the Union and the Confederacy made attempts (reminiscent of General Hsin's) to launch hot air balloons laden with incendiary devices, hoping that they would drift over enemy lines and explode on impact with their enemy's ammunition stores, killing nearby troops and destroying critical infrastructures.[35] In 1895, soldiers of one of the US Army's infantry regiments were attempting to find an aerial surveillance solution that did not involve placing a human in an elevated and open position. Their solution was to rig a camera with a timing device that was then raised with a balloon, allowing them to take wide focus pictures of their surroundings.[36] Alexander Graham Bell followed shortly thereafter with the telephone, which was quickly followed by the first long-distance wireless radio transmissions.[37] By 1897, Nikola Tesla had built and demonstrated the first radio-guided electronic torpedo, designed to inflict catastrophic damage on enemy ships at minimal risk to the user.[38] Interestingly, the US Navy disliked this system and was of the opinion that the sneak attacks which it enabled were 'less-than-honourable' and not in keeping with the naval traditions of the time, forcing Tesla to find less scrupulous buyers abroad.[39]

Notwithstanding the importance of the above developments, there is a reason for retrospectively describing World War I as the dawn of current forms of unmanned systems: it was only at this time that they began to show the true characteristics of present day systems, in both a physical and operational sense. Following on from Tesla's work, the Germans built over a dozen large gasoline powered teleoperated motorboats for use against British targets.[40] These carried explosives that detonated on impact and were initially tethered to a land-based control tower and, later, an overflying seaplane. Later again, they were able to be radio-controlled. However, there were difficulties coordinating the warship, seaplane and land station. These difficulties, coupled with engine troubles, resulted in a lack of confidence surrounding the use of these unmanned vessels, with Germany instead displaying preference for the well-tested *U-boat*.[41] Toward the end of the War, an American engineer with the Caterpillar Tractor Company designed one of the earliest teleoperated unmanned ground vehicles called the *Land Torpedo*.[42] From patents, it is known that the design of this tracked and tethered vehicle was motivated by a desire to avoid the horrors of early trench warfare and was thus designed to deliver communications cables from trench to trench. It was also capable of levelling terrain between trenches prior to infantry advances, or delivering explosives directly to enemy trenches.[43] While a similar design was put to use in Italy, these platforms came too late for use in World War I.

The development of unmanned aerial vehicles began much earlier, primarily due to the heavy casualties that Britain encountered as a result of early German warplanes. Elmer Sperry made one of the most important aeronautical developments in 1915 with the creation of a levelling device known as the gyrostabiliser, which is at the heart of most modern autopilots and many modern robots.[44] Later, in 1917, the US Navy gave Sperry's son a grant to build an 'aerial torpedo', with the name again indicating the fact that they were designed to deliver weapons which would detonate on impact, destroying the delivery vehicle.[45] This led to the development of the Curtiss *N-9* catapult-launched seaplane, which was capable of basic self-guided flight, but had a pilot to carry out the more complex take-off. It was a flight of this kind that claimed the life of Joseph Kennedy, elder brother of the future US President, John F. Kennedy. A radio-controlled version was later made, but it was just as crude, with its analogue instruments not allowing precision flight or weapons delivery.[46] The US Army designed a similar aerial torpedo in 1918, named the *Kettering Bug*.[47] The British also had several programs underway, mostly based on existing aircraft frames retrofitted with expendable engines. All of these systems were rather unreliable. This included a DeHavilland model that reportedly crashed into a crowd of civilian spectators.[48]

Unmanned aerial vehicle development continued throughout the *interbellum* period, driven primarily by the British, despite their earlier failures. This persistence is attributable to doubts within the British Armed Forces concerning whether the Royal Navy's surface fleet was capable of surviving an aerial attack.[49] Because of this concern, the British dedicated much of their time to developing 'aerial targets' which were, in essence, practice targets for the Navy's anti-aircraft gunners. World War II constituted a slower period for development, though there were some notable achievements. In 1939, the Borgward Company of Germany, drawing on a French prototype, created a small electric tracked demolition vehicle with a container for explosives. The teleoperated machine would be directed to its position by radio control. Coupled with a timing device, the explosives would then be dropped and the machine would retreat, making it one of the first recoverable robots of its sort (although, in practice, many failed to retreat in time).[50] In the 1940s, the British firm W.H. Allen & Company also began researching the options for an enormous thousand-tonne legged armoured tank, but funding was soon withdrawn for more urgent needs. At this time, there was little other interest in deploying unmanned platforms among Western nations.[51] Taking advantage of this, toward the end of World War II the Germans introduced the *Fiesler 103*, also known colloquially as the 'V1 Buzz Bomb' because of the buzzing sound it made when descending to its target. The V1 – a precursor to later missiles – was cheap and simple, and with the aid of a reliable pulsejet engine, could carry a large payload over a range of a few hundred kilometers.[52] Thousands of these V1s were used to great effect in the terror bombing of London and southeast England, instilling fear in the population and establishing the infamous reputation of this weapon.[53]

Compared to the pace of earlier innovation, the Cold War period brought a comparatively rapid development of all things unmanned. There was great concern about intercontinental ballistic missiles during the Soviet era, but most nations also recognised the ongoing need for other innovative technologies and many important developments were made during the first half of this period. In the 1950s and '60s, both General Electric (GE) and Ohio State University (OSU) had interest in legged robots, developing manned prototypes that would be able to cross the most difficult terrain while carrying loads of equipment. GE's *Cybernetic Walking Machine* was a quadruped, while OSU's *Adaptive Suspension Vehicle* was a more complex hexapod, which mimicked the tripedal gait of six legged insects (with robotic legs touching the ground in alternating triangles, much like a cockroach).[54] In 1953, the US built the *Cable-Controlled Underwater Recovery Vehicle* or *CURV*, a teleoperated unmanned submarine capable of performing inspection, survey, retrieval and disposal tasks, all via a display screen, removing humans from the hazardous underwater environment.[55] Both the French and Russians later designed similar systems with improved operator interaction systems, sensors and robotic arms for the improved manipulation of objects.[56] During this period, the public also got their first view of an advanced and recoverable unmanned aerial vehicle with the introduction of the Ryan *Firebee* by the US (some of which were given to Israel).[57] The *Firebee* was one of the most widely used drones, with use beginning in the 1950s and newer models extending right through to the present day. Many variants were weaponised, but with the loss of manned *U2* reconnaissance aircraft over Soviet Russia and Cuba in the 1960s, some variants were later fitted with high quality cameras and electronic sensors to perform politically sensitive reconnaissance missions.[58] These recoverable reconnaissance drones were launched from under the wing of a manned *DC-130* aircraft. Once released, they would fly their mission via wireless link and then ditch in the ocean, after which they would be pulled out of the water and prepared for their next mission.[59]

The latter half of the Cold War period saw the continued drive for greater technologies, owing to concern about the threat posed by ballistic missiles and the use of technology to explore outer space. Nations around the globe continued to purchase unmanned systems. The British bought the *Phoenix* reconnaissance system, which could relay reliable near-real-time images and data.[60] The US Navy and its Marine Corps took delivery of the *Pioneer*, which could be teleoperated from a system small enough to be placed in a soldier's backpack.[61] Various vertical take-off and landing systems (commonly rotorcraft) and aerostats (blimps) were also developed. Unmanned submarines were pushed to new limits, with those like the American *Deep Drone* being used for a variety of tasks at depths well in excess of a kilometre below sea level.[62] It was also around this time that unmanned vessels began successful untethered operation with improved acoustic, radio and laser communications. This was alongside investigations into supervised autonomous operation. Unmanned fast boats were put to effective use as well, with those like the *Seaflash 300* – a teleoperated

Jet Ski-like vessel controlled by long-range radio and capable of reaching speeds of over 25 knots – used by the Royal Navy and other European naval forces.[63] Furthermore, the then-young US Advanced Research Projects Agency (later renamed the Defense Advanced Research Projects Agency or DARPA) advocated the development of unmanned frigates and destroyers with high-level autonomy. Unmanned ground vehicles also became much more sophisticated. With the creation of more refined communication links, displays, controls and actuators, many small teleoperated robots were built, some with rotating robotic arms, microphones, cameras, shotguns and grenade launchers. Larger systems remained important, though they were not widely deployed. Robot Defense Systems engineered the *Prowler*, an armed terrestrial vehicle the size of a small truck. Performing well when teleoperated, it proved less robust when under supervised autonomous operation, limited by early microprocessors, range finders and radar.[64] It was also in this period that DARPA began to design more advanced ground vehicles with task-level autonomy. To achieve this, DARPA partnered with academic institutions and this saw the creation of robotics labs at the Massachusetts Institute of Technology (MIT) and Stanford University, together cultivating the development of many systems in operation today.

The Twenty-first Century: The Era of Task-level Autonomy

As evidenced by the discussion so far, the United States has been the most active in the development of unmanned systems. With this remaining true today, the discussion of present-day developments will be necessarily American-focused.[65] Also, rather than taking a roughly chronological approach, this section will look at the systems currently in use across each domain – land, sea and air – separately. The aim here is to provide a very clear picture of the systems that are the focus of the remaining chapters.

Land Systems

Modern unmanned ground vehicles can be grouped into two general categories that have developed alongside the technology: there are large versions like armoured combat and support vehicles that tend to be capable of task-level autonomous operation; and there are others which are comparatively small and tend to be non-autonomous or need supervision, such as explosive ordinance disposal robots. Alongside these, there are less-common anthropomorphic systems designed for shipboard firefighting and the like. Many of the larger vehicles draw on technologies developed for the DARPA Grand Challenge competitions that were held in 2004/2005 and offered significant prize money of around $1 million for the first three vehicles that could navigate a harsh

140-mile long track through the Mojave Desert without human supervision.[66] One such derivative vehicle is the *Crusher*, a joint effort between DARPA and Carnegie Mellon University, designed to perform fire support, reconnaissance, logistics, sentry or medical rescue roles using modern processors, lasers and a Global Positioning System (GPS).[67] The *Crusher* is capable of teleoperation, waypoint navigation or limited task-level autonomous operation across steep inclines, creek beds and rough terrain.[68] Another good example is the *Guardium*, produced by a conglomerate of Israeli manufacturers. Unlike the expensive tank-like *Crusher*, the *Guardium* is a cost effective, lightly armoured, high-speed vehicle capable of on – and off-road operation with limited task-level autonomy. It has sophisticated sensors capable of identifying intruders within a defined proximity, who can then be engaged by remotely controlled weapons.[69] Boston Dynamics' *BigDog* is also worthy of mention as a prototype quadruped pack mule about the size and shape of a donkey. It is designed to accompany soldiers on foot under real operational conditions, without need for regular input. Unlike earlier unsuccessful quadrupeds, the self-contained *BigDog* can traverse difficult terrain with a load of up to 154 kilograms, all while displaying better-than-human balance and stability.[70] Boston Dynamics also manufacture the *RiSE*, a system that can traverse walls, windows and other vertical terrain much like a spider, potentially recording valuable covert intelligence as it proceeds.[71]

Equally important are the smaller sized robots. Among them, systems like Recon Robotics' *Throwbot* and Omnitech's *Toughbot*.[72] Both are designed in the shape of a small cylinder with wheels on either end so that they can be tossed down a corridor, up a stairwell or into a window and operate no matter how they land. These durable battery-operated reconnaissance robots are teleoperated and allow troops to search for and assess any traps, explosive devices or enemy personnel. These unmanned vehicles, as useful as they are, are not able to alter their environment. For this, troops must turn to other robots like the US Marine Corp's *Dragon Runner*. This vehicle is slightly larger and has a 'dump body attachment', allowing it to deliver charges to destroy improvised explosive devices in addition to performing normal route clearing functions.[73] It is manipulated by something akin to a Sony PlayStation controller, which the Navy hopes will assist young marines in their training efforts.[74] Slightly larger again, there is the *Packbot* from iRobot and the *Modular Advanced Armed Robotic System* or *MAARS* from Foster-Miller. Both of these robotic ground vehicles have been used extensively in Iraq and Afghanistan. Most, if not all, run on tracks, but some are also equipped with 'flippers' that allow the robot to climb stairs or to overcome other obstacles.[75] With sophisticated claw-like grippers, high-powered zoom cameras and nuclear, biological and chemical sensors, the overwhelming majority of these systems are employed in explosive ordinance capacities.[76] Others are intended for combat support and can be fitted with a wide range of weapons, from less-than-lethal rubber pellet shotguns to deadly machine guns and rocket launchers. These systems are generally teleoperated not too far from the conflict zone, but newer models can make their way to predesignated points without input from an operator.

Maritime Systems

New concerns over the survivability of maritime assets have also driven the twenty-first century development of sea-going systems. During the interwar period, the concern was over aerial attacks, but in 2000, 17 sailors were killed and many more injured when a small surface vessel approached the US Navy destroyer *Cole* and exploded on impact with its hull.[77] This highlighted the need for small swift-moving force protection vessels. In response, the Navy developed the *Spartan Scout*. It resembles a rigid hull inflatable boat and is typically operated by remote control.[78] Another vehicle developed in response to the emerging terrorist threat is the task-autonomous *Protector*, developed by Israeli Rafael Advanced Defense Systems and made available to Britain, Singapore and the US. The *Protector's* V-shaped hull and streamlined design allows it to reach speeds of up to 50 knots and to monitor seas for nearly a day. It is equipped with a stabilised machine gun for interception, sophisticated electro-optical sensors for day or night surveillance and a public address system to warn nearby vessels of their unwanted presence. More recently, systems have delivered even greater endurance. Zyvex Technologies have recently field-tested a large long-range unmanned vessel called the *Piranha*, aimed at fulfilling patrol and reconnaissance roles.[79] It is constructed of a nano-enhanced carbon fibre, making the vessel lighter and tougher than others. There is also one small vessel that is supplemented by solar power, improving capability and reducing environmental impact.[80]

Modern unmanned submarines have continued on what is a typical trajectory across all the domains: shifting away from teleoperation and becoming increasingly autonomous. This is partly because today's communications technologies continue to perform poorly at distance under water. In recent years, several systems have reached levels of maturity making them viable options for operation in the field. For example, the *Long-Range Mine Reconnaissance System* is capable of being launched, and recovered, from the torpedo tube of a slow moving submerged submarine in much the same way that Special Forces divers are released. Using advanced processors and sensors, it is also capable of extended task-autonomous operation, allowing for missions of up to five days in length. The main aim of this system is to provide submarine crews with clandestine mine reconnaissance, while freeing up sailors to perform other work.[81] With underwater technologies like this only recently reaching maturity, tactical applications are currently limited. However, British Aerospace Engineering's futuristic *Talisman* seems to represent the next step in unmanned submarines and is rapidly evolving. It is constructed of an aerodynamic carbon-fibre body and is fitted with a number of vector thrusters like modern cruise ships, providing precision movement in all directions. Mission parameters are set prior to launch for task-autonomous operation but, if needed, can be reprogrammed via an acoustic communication system while under water or by radio/satellite when surfaced.[82] The *Talisman* has a multi-role capability, with different configurations able to detect and neutralise mines, perform littoral reconnaissance as well as fire and control weapons.

Aerial Systems

Following on from the earlier developed *Pioneer*, which saw successful operation in the Gulf War, aerial drones became one of the most exciting areas of technological advance within the US military. Today, there are dozens of different platforms in use across the various services. Among the most widely used is Northrop Grumman's *RQ-4 Global Hawk*, which is a dedicated surveillance aircraft.[83] Made from composite materials, it is capable of task-autonomous extended-range high-altitude reconnaissance, providing commanders with near-real-time high-resolution intelligence.[84] More commonly seen in the media is the General Atomics *RQ-1* or *MQ-1 Predator*.[85] The *Predator* is primarily a middle-range reconnaissance and target acquisition vehicle. In place of windows, life support, ejection systems and so on, it carries sophisticated cameras and radars for viewing through atmospheric haze and a laser for designating targets. The *Predator* is typically launched from a normal runway and can be supervised or controlled from afar with satellite communications and a control pod resembling a cockpit.[86] It was used in combat over Afghanistan, Bosnia, Iraq, Libya, Pakistan, Serbia, Somalia and Yemen, among other places. The *MQ-9 Reaper*, a derivative, can fly twice as high and as fast, but importantly, has been adapted to carry an extensive weapons arsenal, making it an effective 'hunter-killer'.[87] Furthermore, both Boeing and Northrop Grumman have demonstrated drones designed specifically for stealth combat and aerial bombing. The manned *Joint Strike Fighter*, funded by Australia, Canada, Denmark, Italy, the Netherlands, Norway, Turkey, the US and the UK, has also been designed to accommodate the potential need to convert it into an unmanned aircraft at a later date.[88]

Besides the larger aerial drones, there are many other smaller systems with different uses. Among them is the *MQ-8 Fire Scout*, designed for the US Navy. It is a smaller-than-normal unmanned helicopter that rivals the specifications of many full-sized helicopters, but is also task-level autonomous, being able to carry out independent shipboard operations and flight.[89] That is, once commanded, the automated *Fire Scout* can take off, fly a certain route, and return to the helicopter deck with a smooth landing, despite the yaw, pitch and roll of a ship and without continuous input. The *Fire Scout* is ideally suited to over-the-horizon reconnaissance and surveillance, but is also a candidate for weaponisation or re-supply uses. As well, there are many typical winged drones that have been reduced in size for transportation with expeditionary troops. One Californian company, AeroVironment, is perhaps most well-known for the development of these small systems. They produce the semi-autonomous *Dragon Eye*, which has a limited range but is unique in that it can be launched by a bungee system. Coupled with a colour or infrared camera, it is used for tactical (or over-the-next-hill) reconnaissance by the Marine Corps.[90] AeroVironment is also responsible for producing a number of micro-aerial vehicles, including the *Black Widow* at a mere 15.2 centimetres wide and 80 grams in weight, capable of flying for 30 minutes while providing a video downlink to its operator.[91] The recently produced *Nano Hummingbird* is another remote-controlled micro-drone, built to resemble and fly like a hummingbird. Equipped with a small camera, it has proved relatively successful in early flights.[92] Both of these tiny unmanned aerial vehicles are designed primarily for urban reconnaissance and surveillance purposes.

Toward an Autonomous Future

Thus far, this chapter has outlined the gradual rise of unmanned systems. Yet with the pace at which these systems are developing, it is particularly important to look forward and anticipate *future* directions and change. However, just as it is difficult to predict whether we are on the cusp of a new revolution in military affairs, making anything but near-term predictions about the technology often proves problematic because of the inductive reasoning we use to predict the future.[93] That said, reasonable estimation dictates that the twenty-first will be another bloody century with evolving strategies and technologies.[94] There is also much that can be garnered from the general trends associated with these systems while leaving the precise 'point predictions'[95] to futurologists such as Ray Kurzweil and the wide range of lobby groups and corporations that stand to benefit from the growth of the unmanned systems industry and the wider military-industrial complex.

One thing that seems particularly clear is that unmanned systems are likely to continue progressing up the scale of so-called 'autonomous' operation. As has been outlined in this chapter, most early unmanned systems were teleoperated. By World War II, the systems in use had already evolved to include supervisory and basic task-level autonomy. One technological reason behind the early drive for autonomy had to do with communications. It was often difficult to send signals to systems at great depths, heights or over rugged terrain, and those that were sent were vulnerable to interception and attack.[96] The more autonomous the system, the fewer the control signals needed. Today, many systems are capable of advanced task-level autonomy (even though bandwidth and interception are persistent problems), but the decision to use lethal force generally rests with a human operator, although some systems are capable of firing their weapons at predesignated targets. However, in the future we may well see the removal of human beings from the decision-making loop. That is, with software and computer processors becoming ever more advanced, it *might* be possible to create fully autonomous robots capable of learning and/or determining under which circumstances it should use lethal force. Ronald Arkin, for instance, has detailed something he calls the 'ethical governor', which is essentially a software and hardware package that would guide and restrict robots in the application of lethal force, in a manner he thinks consistent with the relevant laws of war and rules of engagement.[97] According to Arkin, a fully autonomous system would generate a tactical option that would then be filtered through the governor and be suppressed if it failed to meet the logic of the system (but this optimism will be critiqued in subsequent chapters).

Despite issues with bandwidth and the like, it seems relatively likely that in the near future unmanned systems will begin to be networked together as part of a larger network-centric operations system. The need for networked operations was already evident in World War I. Recall that the Germans had problems coordinating their early tethered boats. Indeed, the coordination between warship, seaplane and land station was abysmal. If the operation of the three were linked together, the fate of the boat program may have been different. With computers, today's planes, ships and soldiers no longer have to communicate by carrier pigeon, semaphore signals or radio, but there is more to come. The future holds promise for enhanced levels of connectivity and information sharing, which would speed up operations and might make those involved more battlespace aware. A network properly linked together will usually be 'smarter' than the sum of its individual components.[98] As a part of this networked system, it is likely that different unmanned platforms will come to operate in 'swarms' of increasing size and complexity, and do so much effectively than current platforms. A swarm, in the context of robotics, involves a number of robotic artefacts coming together to operate as a whole.[99] Much like a swarm of bees or a flight of birds, there is no single controller of a swarm. Instead, each robotic artefact is linked together with some other/s to create a sub-network that overcomes some communications problems and allows the swarm to act in a self-organising unison to reach a common goal. In terms of unmanned systems, this could mean that each platform would be given a strategic goal and sent on its way, collaborating with other platforms in the swarm to achieve the necessary operational task/s. Of course, the issue here is the unpredictability of swarming systems. Unpredictability might be a positive thing with which to confront one's enemy, but in terms of command and control it is of potential moral concern. For the moment, at least, there is no need for worry as genuine swarming is not possible. What we will soon come to see more of, as a precursor, is that human operators will be responsible for overseeing multiple-robot systems.[100] The software systems they use for this may eventually be made 'open source', which would introduce a range of other problems.

In terms of physical design, future unmanned systems are likely to take a form that supports their collaborative role in swarms and the need to operate across a variety of different domains while minimising complexity and maintenance costs. In other words, they will need to be capable of physically interacting with each other to perform across land, sea and air, or be able to do so on their own. Initially, as is already happening with aerial drones, this might mean that systems across the different domains are built on a common platform, with add-on mission packages. It might also mean that some platforms are designed to be integrative. For example, a number of unmanned ground vehicles may come together, interlock themselves and create a landing space for an unmanned helicopter. There may also be developments evolving out of today's primitive self-reconfiguring robots that can take on various forms and functions to suit their environment. These are often made of many small and identical modules that come together to form a larger robot or split to create many smaller robots. If these modules can be created at the nanoscale level, a self-reconfiguring robot could take on almost any form. DARPA also has a chemical robot program running. Under this 'Chembot' program, MIT and Boston Dynamics are developing the *SQUISHBot*, a tiny robot that is able to move like a slug by secreting a viscous fluid and morph its visco-elastic skeleton to fit into small holes and cracks, possibly to create seals or otherwise gain access to sensitive areas.[101] On the general topic of size, those unmanned systems that need to be small, such as those used in covert operations, will continue to grow smaller with the miniaturisation of the technologies within them.[102] With the aid of new power sources and fuel cells, many unmanned systems will also benefit from drastically improved endurance, to the point where persistent or 'around the clock' reconnaissance operations are possible.

Conclusion

Humans have always sought new weapons technologies with which to fight their enemies more effectively and efficiently, and the widespread development and deployment of unmanned systems that is currently underway follows much the same pattern that has existed for thousands of years. These systems have their roots in ancient mythology, according to which the Greek gods and ancient Chinese generals benefited from their use. Apart from some theoretical developments by those like da Vinci and others in the early nineteenth century, their first genuine operation occurred in World War I, with systems that were generally tethered to a human operator and destroyed upon impact with their targets. World War II resulted in the extension of these systems and the introduction of early recoverable unmanned ground vehicles. Armed forces were slow to adapt to these systems as many were unreliable, with engine and control link failures being regular occurrences. Therefore, the preference arose for more reliable and effective manned systems. This was the case until the Cold War period, when radio-controlled recoverable systems matured, motivated by the ballistic missile and space threat. Today, unmanned systems are almost as interesting as those conceived of by the ancients, thanks largely to the power of modern computer processors. Future systems are likely to be smarter and more autonomous, be capable of operation in swarms and will in many cases be smaller, more durable and able to operate across a wider variety of domains than is currently the case. Taken together, the innovations that characterise the current revolution represent a more pernicious threat than any weapons system created to date, with humanity only recently acquiring the option of deploying weapons that so greatly mediate the operator's role and may ultimately take action independently of their human users. Therefore, the justifications for employing these weapons must be carefully considered if the various challenges are to be managed or converted into practical and beneficial solutions, and this task is taken up in the next chapter.

Notes

1 The first proper war is thought to have taken place between Sumer and Elam (present day Iraq and Iran) in 2700 BC. See Richard A. Gabriel and Karen S. Metz, *From Sumer to Rome: The Military Capabilities of Ancient Times* (New York: Greenwood Press, 1991), 4–7.

2 For a more complete treatment of this topic, see Jai C. Galliott, ed. *Commercial Space Exploration: Ethics, Policy and Governance* (Farnham: Ashgate, forthcoming).

3 Yuri Balyko, 'NATO's Unmanned Aerial Vehicles in Local Conflicts', *Military Parade* 1 (2008): 36.

4 Johannes Kuschel, *The European Defense Market: Great Challenges – Small Moves* (Norderstedt: GRIN Verlag, 2008), 108; Kenzo Nonami et al., *Autonomous Flying Robots* (Dordrecht: Springer, 2010), 6.

5 United States Department of Defense, 'FY2009-2034 Unmanned Systems Integrated Roadmap' (Washington, DC: Department of Defense, 2009), 4.

6 Ibid., 5.

7 Michael O'Hanlon and Peter W. Singer, 'The Real Defense Budget Questions', *Politico*, 21 July 2011, 2; Peter W. Singer, 'U-Turn: Unmanned Systems Could Be Casualties of Budget Pressures', *Armed Forces Journal* (2011); Paul L. Francis, 'Issues

to Be Considered for Army's Modernization of Combat Systems', ed. United States Government Accountability Office (Washington, DC: United States Government Accountability Office, 2009).

8 Office of the Under Secretary of Defense (Comptroller), *Program Acquisition Cost by Weapon System* (Washington, DC: Department of Defense, 2013).

9 Ibid.

10 United States Environmental Protection Agency, 'EPA's Budget and Spending', http://www.epa.gov/planandbudget/budget.html.

11 Peter M. Asaro, 'Military Robots and Just War Theory', in *Ethical and Legal Aspects of Unmanned Systems: Interviews*, ed. Gerhard Dabringer (Vienna: Institute for Religion and Peace, 2010), 105.

12 Steven M. Shaker and Alan R. Wise, *War without Men: Robots on the Future Battlefield* (Washington: Pergamon-Brassey's, 1988), 13.

13 Daniel W. Stuckart and Michael J. Berson, 'Artificial Intelligence in the Social Studies', in *Research on Technology in Social Studies Education*, ed. John Lee and Adam M. Friedman (Charlotte: Information Age Publishing, 2009), 231.

14 Homer, *The Iliad and Odyssey of Homer*, trans. Alexander Pope (London: George Routledge and Sons, 1890). This translation is probably best for those already relatively familiar with the Iliad, while Lattimore's much more modern translation (cited below) is more suitable for novices.

15 Sidney Perkowitz, *Digital People: From Bionic Humans to Androids* (Washington, DC: Joseph Henry Press, 2005). 51–84.

16 Homer, *The Iliad of Homer*, trans. Richard Lattimore (Chicago: University of Chicago Press, 1951), 18.136.

17 Ibid.

18 For an authoritative treatment, see Aimee van Wynsberghe, *Care Robots: An Ethical Framework* (Farnham: Ashgate, forthcoming).

19 M. E. Moran, 'The History of Robotic Surgery', in *Robotics in Genito-Urinary Surgery*, ed. Ashok Kumar Hemal and Mani Menon (Dordrecht: Springer, 2010), 9.

20 The 300 BC silver Didrachm of Phaistos (an ancient coin) depicts a winged Talos, holding a rock in a defensive posture.

21 Pseudo-Apollodorus, *The Library of Greek Mythology*, trans. Keith Aldrich (Lawrence: Coronado Press, 1975), Bibliotheca 1.27.

22 *Greek Elegiac Poetry: From the Seventh to the Fifth Centuries B.C.*, trans. Douglas E. Gerber (Cambridge: Harvard University Press, 1999), Mimnermus, Fragment 12.

23 Pseudo-Hyginus, *The Myths of Hyginus*, trans. Mary A. Grant (Lawrence: University of Kansas Press, 1960), Astronomica 2.15.

24 Homer, *The Odyssey*, trans. Walter Shewring (Oxford: Oxford University Press, 1980), 7.93.

25 Pseudo-Apollodorus, *The Library of Greek Mythology*, 1.127.

26 Joseph Needham, *Science & Civilisation in China: History of Scientific Thought*, trans. Ling Wang (Cambridge: Cambridge University Press, 1956), 53.

27 *The Mahabharata*, trans. Protap Chandra Roy (Calcutta: Bharata Press).

28 R. Clark, *The Role of the Bomber* (New York: Thomas Y. Cromwell Company, 1977), 9.

29 Joseph Needham, *Science and Civilisation in China: Physics and Physical Technology* (Cambridge: Cambridge University Press, 1965), 577.

30 Joseph Needham, *Science and Civilisation in China: Chemistry and Chemical Technology* (Cambridge: Cambridge University Press, 1987), 157–58.

31 Moran, 'The History of Robotic Surgery', 11.

32 Noel Sharkey, 'I Ropebot', *New Scientist* 194, no. 2611 (2007): 3.

33 Mark E. Rosheim, *Leonardo's Lost Robots* (Dordrecht: Springer, 2006), 105–13.

34 Walter J. Boyne, *How the Helicopter Changed Modern Warfare* (Gretna: Pelican Publishing, 2011), 301.

35 James Crabtree, *On Air Defense* (Westport: Greenwood Publishing, 1994), 2.

36 J. Finnegan, *Military Intelligence: A Picture History* (Arlington: United States Army Intelligence and Security Command, 1985), 10.

37 Springer, *Military Robots and Drones.*

38 Krishnan, *Killer Robots: Legality and Ethicality of Autonomous Weapons*, 15; Michael Burgan, *Nikola Tesla: Physicist, Inventor, Electrical Engineer* (Mankakto: Compass Point Books, 2009), 68.

39 Springer, *Military Robots and Drones*, 8–9.

40 Gordon Williamson, *German E-Boats, 1939–45* (University Park: Osprey Publishing, 2002).

41 Ibid., 3–4.

42 Unknown, 'Who Called Them "New"?', *Popular Science*, June 1945, 78.

43 Elmer E. Wickersham. Land Torpedo. United States Patent 1407969, filed 28 February 1922.

44 Shaker and Wise, *War without Men: Robots on the Future Battlefield*, 22.

45 Ibid.

46 Laurence R. Newcome, *Unmanned Aviation: A Brief History of Unmanned Aerial Vehicles* (Reston: American Institute of Aeronautics and Astronautics, 2004), 16–21.

47 Shaker and Wise, *War without Men: Robots on the Future Battlefield*, 21.

48 Ibid.

49 Joseph Moretz, *The Royal Navy and the Capital Ship in the Interwar Period: An Operational Perspective* (London: Frank Cass Publishers, 2002), 128–29.

50 Jean-Denis Lepage, *German Military Vehicles of World War II: An Illustrated Guide to Cars, Trucks, Half-Tracks, Motorcycles, Amphibious Vehicles and Others* (Jefferson: McFarland & Company, 2007), 164–66.

51 Krishnan, *Killer Robots: Legality and Ethicality of Autonomous Weapons*, 19.

52 Steven J. Zaloga, *V-1 Flying Bomb 1942–52: Hitler's Infamous 'Doodlebug'* (University Park: Osprey Publishing, 2005), 3.

53 The V1s were not technically efficient, as many failed to reach their designated targets, but they were effective in the sense that they fulfilled their more strategic goal of instilling fear in the population.

54 Daniel V. Hunt, *Smart Robots: A Handbook of Intelligent Robot Systems* (Dordrecht: Springer, 1985), 26.

55 Robert F. Marx, *The History of Underwater Exploration* (Mineola: Dover Publications, 1990), 188.

56 Ibid.

57 Steven J. Zaloga, *Unmanned Aerial Vehicles: Robotic Air Warfare 1917–2007* (Oxford: Osprey Publishing, 2008), 12.

58 Ibid.

59 Ibid.

60 Shaker and Wise, *War without Men: Robots on the Future Battlefield*, 94.

61 Ibid.

62 Richard Bartlett, *United States Navy* (London: Heinemann Library, 2004), 25.

63 Eric Wertheim, *Naval Institute Guide to Combat Fleets of the World: Their Ships, Aircraft, and Systems* (Annapolis: Naval Institute Press, 2007), 511.

64 Richard S. Friedman, *Advanced Technology Warfare: A Detailed Study of the Latest Weapons and Techniques for Warfare Today and into the 21st Century* (New York: Harmony Books, 1985), 132.

65 The United States has historically taken a protectionist stance with regard to industry and trade, and, with military technology supporting America's superpower status, there is an obvious motivation for keeping important developments in-house or restricted to close allied nations. However, unless the United States relaxes its extremely tight arms export controls, it is likely that other nations will be forced to continue developing their own unmanned systems due to a lack of commercially available systems. Eventually, these nations may surpass the United States in unmanned systems production, particularly once Coalition operations in Iraq and Afghanistan draw to a full and proper close. This proliferation is likely to exacerbate some problems and alleviate others.

66 Deborah D. Stine, *Federally Funded Innovation Inducement Prizes* (Darby: DIANE Publishing, 2009), 8.

67 United States Department of Defense, 'FY2009-2034 Unmanned Systems Integrated Roadmap', 112.

68 Ibid.

69 D. P. Sellers et al., *International Assessment of Unmanned Ground Vehicles* (Alexandria: Institute for Defense Analyses, 2008), S3, 29.

70 Marc Raibert et al., 'Bigdog, the Rough-Terrain Quadruped Robot' (paper presented at the 17th World Congress, Seoul, Korea, 6–11 July 2008).

71 T.L. Lam and Y. Xu, *Tree Climbing Robot: Design, Kinematics and Motion Planning* (Dordrecht: Springer, 2012), 10.

72 Mitchell Barnes, H. R. Everett, and Pavlo Rudakevych, 'Throwbot: Design Considerations for a Man-Portable Throwable Robot' (paper presented at the SPIE, March 2005).

73 United States Department of Defense, 'FY2009-2034 Unmanned Systems Integrated Roadmap', 119.

74 United States National Research Council, Committee on Autonomous Vehicles in Support of Naval Operations, *Autonomous Vehicles in Support of Naval Operations* (Washington, DC: The National Academies Press, 2005), 141.

75 Peter W. Singer, 'Military Robots and the Laws of War', *The New Atlantis: A Journal of Technology and Society* 23 (2009): 33.

76 This is important to note because many people think of these unmanned ground vehicles as 'killer robots' when, in most cases, they are not actually used in a lethal capacity. It should also be noted that their use in other-than-lethal roles is just as important from an ethical perspective, as will be demonstrated in forthcoming chapters.

77 George Lucas, 'The Ethical Challenges of Unmanned Systems', in *Robots on the Battlefield: Contemporary Issues and Implications for the Future*, ed. R. Doare, D. Danet and J. P. Hanon (Fort Leavenworth: Combat Studies Institute Press, 2014), 137.

78 Anthony Finn and Steve Scheding, *Developments and Challenges for Autonomous Unmanned Vehicles: A Compendium* (Berlin: Springer Verlag, 2010), 22.

79 Michael Nemeth, 'Zyvex Technologies Reveals the Piranha Unmanned Surface Vessel', http://www.prweb.com/releases/zyvex/piranhaUSV/prweb4658434.htm.

80 J. Jalbert, 'Solar-Powered Autonomous Underwater Vehicle Development' (paper presented at the 13th International Symposium on Unmanned Untethered Submersible Technology, Durham, August 2003).

81 United States National Research Council, *Autonomous Vehicles in Support of Naval Operations*, 126–27.

82 BAE Systems, 'BAE Systems Unveils Latest Innovative Unmanned Underwater System', http://www.baesystems.com/Newsroom/NewsReleases/autoGen_1072271 14948.html.

83 Northrop Grumman purchased Teledyne Ryan, maker of the earlier mentioned *Firebee*, in 1999.

84 Ray Bonds, *Illustrated Directory of Modern American Weapons* (Minneapolis: Zenith Imprint, 2002), 210; Thomas P. Ehrhard, *Air Force UAVs: The Secret History* (Washington, DC: Mitchell Institute Press, 2010), 54–6.

85 The 'R' designates a reconnaissance role, while 'M' designates systems with a multi-role function. The 'R' in this name was replaced in 2002, when the system was realised as having multi-role potential. However, as per the earlier footnote on unmanned ground vehicles, unmanned aerial vehicles log many more hours performing reconnaissance than they do tactical support.

86 Anthony Cordesman, *The Iraq War: Strategy, Tactics, and Military Lessons* (Washington, DC: CSIS Press, 2003), 307–8.

87 Fred J. Pushies, *U.S. Air Force Special Ops* (Minneapolis: Zenith Imprint, 2007), 63–4.

88 Thanks to Finney Premkumar of UI Aerospace in Anaheim, California, for bringing this to my attention. Also see Jai C. Galliott, 'Take out the Pilot from Australia's Joint Strike Fighter', *The Conversation* (2014), https://theconversation.com/take-out-the-pilot-from-australias-f-35-joint-strike-fighter-28107.

89 United States Department of Defense, 'FY2009-2034 Unmanned Systems Integrated Roadmap', 66.

90 Ibid., 70.

91 Bonds, *Illustrated Directory of Modern American Weapons*, 208; Jeremiah Hall, 'Low Reynolds Number Aerodynamics for Micro Aerial Vehicles' (University of Colorado at Boulder, 2007), 3–4.

92 W. J. Hennigan, 'It's a Bird! It's a Spy! It's Both', *Los Angeles Times*, http://articles.latimes.com/2011/feb/17/business/la-fi-hummingbird-drone-20110217.

93 Coker, *Ethics and War in the 21st Century*, 173.

94 Colin S. Gray, 'The 21st Century Security Environment and the Future of War', *Parameters: The US Army War College Quarterly Senior Professional Journal* 38, no. 4 (2008): 17.

95 Gary A. Ackerman, 'It Is Hard to Predict the Future: The Evolving Nature of Threats and Vulnerabilities', *Revue Scientifique et Technique International Office of Epizootics* 25, no. 1 (2006): 354.

96 Thomas Adams, 'Future Warfare and the Decline of Human Decisionmaking', *Parameters: The US Army War College Quarterly Senior Professional Journal* 31, no. 4 (2001): 64.

97 Ronald Arkin, 'Ethical Robots in Warfare', *IEEE Technology and Society* 28, no. 1 (2009); Ronald Arkin, *Governing Lethal Behavior in Autonomous Robots* (Boca Raton: CRC Press, 2009); Ronald Arkin, 'The Case for Ethical Autonomy in Unmanned Systems', *Journal of Military Ethics* 9, no. 4 (2010); Ronald Arkin, 'Governing Lethal Behaviour', in *Ethical and Legal Aspects of Unmanned Systems: Interviews*, ed. Gerhard Dabringer (Vienna: Institute for Religion and Peace, 2010).

98 Singer, *Wired for War: The Robotics Revolution and Conflict in the 21st Century*, 184.

99 Mark Coeckelbergh, 'From Killer Machines to Doctrines and Swarms, or Why Ethics of Military Robotics Is Not (Necessarily) About Robots', *Philosophy & Technology* 24, no. 3 (2011): 274.

100 For a number of multi-role control models: see Michael A. Goodrick, 'On Maximising Fan-Out: Towards Controlling Multiple Unmanned Vehicles', in *Human-Robot Interactions in Future Military Operations*, ed. Michael Barnes and Florian Jentsch (Farnham: Ashgate, 2011), 375–80.

101 Jeffrey W. Morin, *Design, Fabrication and Mechanical Optimization of Multi-Scale Anisotrophic Feet for Terrestrial Locomotion* (Massachusetts Institute of Technology, 2011), 17–19.

102 For a good outline of the miniaturisation of technology over time: see Manuel DeLanda, *War in the Age of Intelligent Machines* (New York: Swerve Editions, 1991), 146–54.

Chapter 3
Justifications for the Employment of Unmanned Systems

This chapter explores the moral reasoning behind the use of unmanned systems. The first section will review some of the proposed practical benefits that unmanned systems provide over other weaponry. The second section will examine an account that argues, via a principle of unnecessary risk, that those states with the means to employ these weapons as a force protection measure have a moral obligation to do so. However, it will be argued that this account, with its focus on only one of the benefits of using these systems – saving soldiers – fails to give a complete account of how the obligation to use these systems arises. In order to overcome this, the third section will detail the nature of the relationship between citizen, state and military force. The fourth section will argue that the moral justification for the use of unmanned weaponry is best explained more widely, in terms of a military force's obligation to effectively and efficiently maintain the contract that will be said to exist between it and its state, and indirectly, the people that establish the need for and legitimacy of the state. It follows that if unmanned systems prove ineffective and/or inefficient in particular roles, the justification is limited, as will be claimed in subsequent chapters.

The Operational Case for Unmanned Systems

In making a moral assessment of any technology, new or old, it is necessary to engage in a process of discovering and documenting the potential benefits that are thought to accompany its use, as the justification for the use of a technology will always be closely tied to the advantages it yields. The case is no different with unmanned systems. This is part of the reason why official military policy and strategy documents tend to make clear and early reference to the exploitable advantages associated with the use of unmanned systems technologies.[1] There are many potential benefits for the armed forces and society at large. Some of these will already be obvious from the discussion of the technologies and their uses in the previous chapter. However, at the most general level, the use of these systems is thought to minimise the various costs of war and improve warfighting capability. In this section, some of the specific benefits will be outlined in greater detail before moving on to consider, in the remaining sections, why these benefits are of moral importance, how they are interrelated and why a properly functioning military force ought to seek these sorts of benefits. It should also be noted that, while questions concerning the actual ability of unmanned systems to deliver these benefits, underlying political motivations, and any associated ethical issues are important, they will be left for discussion in later chapters.

The Human Cost of 'Manned' War

One of the inherent benefits of exploiting unmanned systems is that they hold potential to reduce the human cost of war. Traditional military operations typically involve human forces being deployed to remote areas of the world, each with their own unique environmental conditions and geographical features. While there are dangers to be found in any war setting, the most dangerous environmental hazards are those associated with fighting in the common 'hotspots' for human conflict, such as the usually hot, dry and sandy regions of the Middle East and Africa as well as the humid areas of Southwestern Asia and Latin America. Fighting in these climates, especially in the latter ones, also carries with it the additional threat of infectious diseases, a special form of environmental hazard.[2] These natural hazards have a significant impact on deployed forces. Consider the impact of heat alone. During Operation Iraqi Freedom, heat related illnesses took the lives of six soldiers and caused injury to several thousand more.[3] While this is an improvement on the hundreds of British troops that died of heat exhaustion in Iraq during a single month of World War I, these hazards cannot be ignored. Also consider that natural hazards can intentionally be exploited as tools of warfare. For instance, it is estimated that in World War I, upwards of 40,000 soldiers died in the Austrian Alps, not as a result of bullet wounds, but rather of avalanches deliberately triggered by artillery fire.[4] In obvious ways, the use of unmanned systems would minimise a soldier's exposure to these hazards, with the level of mitigation largely dependent on the degree of separation between the operator and their system.

More devastating than the abovementioned natural environmental hazards are those compounding hazards associated with the use of some modern weapons, which have the potential to permanently damage soldiers' health and wellbeing. Bombs, land mines and other indiscriminate weapons frequently result in permanent disfigurement for soldiers who come into contact with them. In 2007, it was reported that nearly 2,000 American soldiers had returned from Iraq or Afghanistan missing at least one limb.[5] The effects of nuclear, biological, chemical and radiological munitions are also of persistent concern. Much like Vietnam veterans who suffer as a result of exposure to Agent Orange, today's soldiers suffer negative health effects resulting from exposure to depleted uranium munitions, amongst others.[6] When these depleted uranium projectiles come in contact with a hard surface, such as armour, cancer-causing radioactive particles are released into the atmosphere.[7] And nanoparticles, used in nano-engineered military technologies, may well present the next challenge.[8] The effects are not yet fully understood, but the concern is that these particles are so small that they may – if disturbed, as is likely in war – enter the body, become lodged and cause damage similar to the impact of asbestos.[9] Again, unmanned systems reduce the risk of soldiers being harmed by distancing them from hazards such as these.

Those who escape physical injury are subsequently confronted with a high chance of developing psychological problems. A United States government funded survey found that two in five soldiers returning from the battlefields of the Middle East often struggle to come to terms with their wartime actions and to readjust to civil society.[10] These soldiers need professional psychological or psychiatric help to overcome anxiety, depression or posttraumatic stress disorders, among others. Christopher Coker writes that some studies report much higher prevalence rates and that, if the United States were to treat all of those affected, the government would be facing not only a mental health crisis, but also a serious financial crisis (additional financial costs will be discussed below).[11] That said, it could be argued that we are already in crisis with over 300 United States military personnel confirmed as having committed suicide in 2009.[12] Either way, the mental health problem is due, in part, to the US Department of Defense's recent invocation of its 'stop-loss' policy – which is authorised by statute and allows military services to retain skilled and experienced personnel by suspending certain laws and policies that prevent retirements and other separations from the military – because it faces/faced a shortage of appropriate enlistees.[13] The use of unmanned weaponry in future conflicts would arguably help soldiers avoid some of the stressors associated with direct combat. As discussed further below, these systems might also reduce workforce requirements and make it easier to train new personnel, which means that these systems would reduce the need to return mentally and physically fatigued personnel to the field or to training positions under the stop-loss policy.

When relying on manned systems, there is also the possibility that troops might be taken captive by enemy forces and exploited. Such events are not uncommon, as was the case with Israeli soldier Gilad Shalit who, in 2011, was released by Hamas after five years in capture.[14] There are a number of concerns here. The first is for the welfare of those who have been captured. Human intelligence is of immense value, so captured personnel are at risk of being tortured (using various deplorable means) for any knowledge they might possess. Of course, it is not just physical harm that is of concern. Those captured and confined will often, similarly to a kidnap victim, suffer psychological stress and trauma resulting from their feelings of powerlessness and betrayal, from the restrictions on their movements and their disorientation in time and space.[15]

The second concern is about the impact on morale and the families of service members. In the past, prisoners of war were paraded publically through the streets to lower the morale of the enemy. Today, in the teleconnected world, captured military personnel are paraded on Al-Jazeera and CNN, rendering the demoralising material readily available for viewing by military personnel and the families of the captured.[16] The third concern is that efforts to rescue captured personnel or those who are vulnerable to such might be quite costly in terms of human life, resources and/or political capital. As in the case of Gilad Shalit, his freedom from enemy forces came at a significant cost to his nation: his release was possible only on the condition that several hundred Palestinian prisoners were exchanged for a single life.[17] The human risk of capture and exploitation is clearly offset by the use of unmanned systems: an opponent who captures an unmanned system will have nothing but the system itself to show for their effort, since the confidential information is usually stored afar, i.e. not in the system, unlike with its human counterpart.[18] Generally, then, unmanned warfare offers soldiers increased protection from a range of physical harms.

Reducing the Financial Cost

Apart from having a tragic human toll, war can also be staggeringly expensive in financial terms. Military budget figures demonstrate that the largest contributor to the overall operational cost of a manned system, accounting for roughly one-third of the total, is the number of personnel directly and indirectly required to operate and support it.[19] Unmanned systems have a great deal of currently unrealised potential to reduce these personnel costs. It is only 'potential' because, as was outlined earlier, unmanned warfare is currently far from 'unmanned'. Operating our current generation of unmanned weaponry remains a labour intensive operation. Take unmanned aerial vehicles. At the moment, a typical platform can be flown with as few as two operators, one controlling flight and the other mission equipment. So in terms of basic workforce requirements,[20] an aerial drone is not all that different from a manned aircraft, with the exception that unmanned vehicle operators working part of a long mission can go home to their families after their shift, whereas traditional aircrew cannot. However, mechanisms for achieving workforce reductions are already in development. Studies have been conducted demonstrating that one operator is capable of flying up to four unmanned aerial vehicles at any given time, so long as the tasks are not overly demanding.[21] And, as demonstrated in the previous chapter, industry is developing more autonomous systems as well as software, which permit operators to spread their attention across a greater number of systems. This means there will be fewer people to equip for traditional battlefield service,[22] drawing fewer military-funded benefits and working fewer paid hours, all of which reduces overall operational costs.

Training is another area in which unmanned weapons would achieve savings in comparison to their manned equivalents. One reason for this is that unmanned systems are generally much more automated than their manned counterparts, meaning that training-hour requirements are lower.[23] This is part of the reason why Predator pilots fly 18 training sorties per year to maintain proficiency while U-2 spy plane pilots fly closer to 100.[24] Unmanned systems further reduce training costs by making it possible to rely more heavily on soft training aids (computer simulations), as the computer game-like operating environment of most unmanned systems is more amenable to replication than those of manned systems. The financial benefit arises because simulation training reduces the need for real-world training sorties, which are expensive and risky. Besides this, unmanned systems designers have purposely designed the controls of unmanned systems to imitate those of gaming consoles such as the Sony PlayStation and Microsoft Xbox. Given that many current and future unmanned systems operators are likely to be familiar with these consoles, they can easily adapt to the control interface, further reducing the number of training hours required.[25] In other words, training costs are reduced through accessing a group of pre-primed operators.

Many unmanned systems technologies are also proving less expensive to procure than similar manned systems. The latter must incorporate engineering provisions to ensure the survival of their human operators. In the case of manned aircraft such as fighter jets and transport planes, these provisions include ejection seats, oxygen tanks, body armour, fire suppression systems and so on. Obviously, there must also be room for the pilots, who occupy space and add further weight to the aircraft. The necessary features of manned systems generally result in them being larger, heavier and more expensive to own and operate than their unmanned equivalents,[26] although they are generally not inexpensive enough to be considered 'disposable'.[27] These systems are still in their technological infancy, so they have not yet demonstrated the downstream cost savings that may eventually be derived from the compatibility and reuse of common parts, training items and broader economies of scale. However even at this early stage, unmanned weapons are often already less expensive than those manned weapons they replace or substitute. For instance, a *Reaper* drone is roughly five times cheaper than an F-22 fighter jet.[28] This will be of greater importance as defence budgets continue to get tighter in North America and Europe and pressure is placed on financial controllers to rationalise defence spending.

Reducing the Environmental Cost

As well as the human and pure financial costs, war also has significant but often ignored environmental costs. The US Department of Defense is the largest single organisational consumer of energy in the world today, consuming almost half a million barrels of oil per day, roughly equivalent to the daily oil consumption of Sweden.[29] Much of this is used to operate the thousands of fossil fuel consuming ships, vehicles and aircraft that it depends on to project force.[30] The issue here, as will be argued later, is not just that the military is depleting the finite reserve of these fuels, but that burning such vast quantities of them has a negative environmental impact. Due to the absence of life support system requirements such as oxygen tanks and ejector seats, unmanned systems are smaller and lighter than manned systems and therefore more fuel-efficient. In addition to their being lighter and smaller, the reduced real world training requirement means that they would consume less fuel overall. Moreover, many of these systems could be powered by renewable solar, wind or vibration energies.[31]

Through the building of infrastructure such as airfields, roads, bridges (which often require the diversion of waterways), temporary accommodation and so on, armed forces often also cause damage to fauna and flora. In like manner, deployed forces can displace farmers from their lands, depriving them of means of self-support and increasing the cost of basic staples for the local populace. 'Scorched earth' strategies – which involve destroying anything that might be useful to the enemy while advancing through or withdrawing from an area – have also resulted in the burning of oil wells and millions of barrels of oil being dumped into the sea, with major ecological (and economic) consequences.[32] Another point worth noting here is that soldiers deployed on foreign soil have often stolen or vandalised precious cultural assets. For instance, Napoleon Bonaparte's army is thought to have caused the irreparable damage to the nose of the Great Sphinx of Giza by using it as a target during weapons practice; while American troops have been accused by the United Nations' cultural agency of damaging the archaeological ruins of Babylon with the movement of their armoured tanks and helicopters, as well as stealing artefacts from various other culturally and environmentally significant Iraqi sites.[33] With unmanned systems reducing the need for troops on the ground, there would clearly be less operational need or opportunity for troops to do any of the damaging and harmful things noted above, thus reducing the military's overall environmental impact.

Enhancing Warfighting Capability

Importantly, at the same time as reducing environmental costs, unmanned systems would enhance the warfighting capability of the armed forces that use them. As mentioned in the previous chapter, these systems would improve access either through their small size or improved manoeuvrability. They can also be sent places that are too dangerous for soldiers. Another advantage mentioned was that these systems, by virtue of not having a human operator on board, would allow for greater endurance and/or persistence, particularly in air operations that have traditionally been limited to relatively short sorties due to concerns about fatigue. Research conducted for the US military also contends that some automated varieties of these systems would, additionally, be more capable than humans, because they can integrate more information from a variety of sources far faster and, among other things, they are not vulnerable to human emotion which can cloud judgement as well as lead to poor operational decision making.[34] Unlike humans, robots do not forget orders, get tired, become afraid, or care whether the soldier next to them has just been shot.[35]

That said, unless unmanned systems lead to a widespread and fundamental change in the character of warfare, military forces will continue to need humans to fight their battles, and these people are hard to attract and retain.[36] The 'stop-lossing' of military personnel, noted above, demonstrates that even those countries that have all-volunteer armed forces sometimes have difficulty maintaining an adequate human resource base to be drawn from in times of need. In resolving this problem, unmanned systems technologies would make it easier for people of lesser physical ability than current requirements to be an asset, particularly among existing enlistees. On top of this, these systems would render women and the disabled eligible for service in some combat roles which may not have been previously open to them. This would not only improve equality by affording members of these minority groups the right to serve their country, but also allow military forces to maintain operational capability.[37] In the next section, we begin to consider the wider role of all of these benefits and place them into philosophical context through the provision of an 'all of things being equal' case for unmanned systems use.

The Ethical Case for Unmanned Systems

Having outlined some of the likely benefits associated with the use of unmanned systems, we now need to consider whether they provide us with a *prima facie* ethical justification for their employment and, if so, how they factor into this justification. For the moment, we will focus on only one of these benefits: reducing the risk of harm to soldiers. This is often cited as the *most* important reason to use unmanned weaponry. This claim is contentious for reasons that will be outlined later, but most would presumably agree that, if the use of unmanned systems can limit the danger faced by soldiers through removing them from harm's way, we would seem to have been provided with good reason using them. Yet few have elaborated on this point and, more generally, it is concerning that the justification is thought to be obvious and uncontentious. For the purpose of this book, it is important to think carefully about the justification and the many ways in which it could be substantiated, as this will influence how we later think about the problems that arise from the use of these systems.

B. J. Strawser is to date the only academic to have provided a detailed account of why we should use unmanned systems.[38] On grounds quite different to those advanced later in this chapter, he contends that not only are we justified in using these systems – that is, their employment is ethically permissible – but that a state with the means to employ unmanned weaponry as a force protection measure is, in fact, ethically obligated to do so. Strawser's contention stems from what he labels the 'principle of unnecessary risk'. This principle holds that if X gives Y an order[39] to carry out some objectively good goal, then X has an obligation, all other things being equal, to select means that make it possible to achieve this goal without violating the demands of justice, making the world worse or exposing the executor to potentially lethal risk, unless the goal cannot be accomplished through other means that are less risky in nature.[40] In other words, it is impermissible to command another to incur any greater potentially lethal risk than is strictly necessary to carry out an action in pursuit of some good goal. Of course, there will be some situations in which the only way to achieve a given good goal is to expose the executor to what is potentially lethal risk. Sometimes this cannot be avoided. Strawser's principle holds only that the least risky alternative must be selected. It does not demand avoidance of all risk.

To try and demonstrate this principle's supposedly modest moral demands and its *prima facie* appeal, Strawser asks us to imagine a scenario involving a group of soldiers fighting against an enemy who is perpetrating some evil that must be halted.[41] The specifics are not important here, but one could easily think of soldiers fighting against a Nazi-like threat. The key point is that the group of soldiers in question is pursuing some objectively good goal. This may include defending their homes and property, their personal wellbeing, the wellbeing of innocents, and so on. At some point during the conflict, the group becomes involved in a fire fight. The officer in charge of the group makes the decision to remove her bulletproof vest, throw down her gun and charge at the enemy with nothing but a rock in her hand. She commands her troops to do the same. Strawser asks us to assume that doing so does not aid the achievement of the goal that they are striving for,[42] yet would dramatically increase the level of lethal risk incurred. He also asks us to set aside the question of whether it is permissible for an officer in charge to do such a thing as an individual, and to ask instead whether it is ethically permissible for the commander to order her troops to lay down their protection and charge the enemy. The principle of unnecessary risk, he says, would rule this out as a moral option, because doing so would involve them facing more risk than is strictly necessary to achieve the desired goal.[43]

Strawser takes this principle of unnecessary risk to be uncontroversial. Its function is to support his main claim, i.e., that if it is possible for a military force pursuing a morally justified action to use unmanned systems in place of manned systems (without incurring any relevant loss of capability), then that military force has an ethical obligation to do so.[44] To further support his claim and, in particular, its normative consequent, Strawser presents us with what is supposed to be a guiding analogy.[45] We have two cities and each has its own bomb squad and highly trained technicians who have generally been successful in disarming and disposing of explosive devices, with few casualties. Both cities have recently acquired unmanned ground vehicles equipped with explosive ordnance disposal tools that allow their technicians to do their job just as well, but without the obvious risk of being killed. One of the cities mandates that the ground vehicles be used whenever possible; the other does not, even though they could easily do so. The latter puts their technicians at risk for no good reason, violating the principle of unnecessary risk.[46] According to Strawser, if it is possible for the bomb squad to use the unmanned vehicle without any loss of bomb-disposal capability, they have an ethical obligation to use it in place of humans. He argues that this case is analogous to that of the military.[47] That is, if one grants that a military force can use unmanned systems equally as well as manned systems, and that this reduces risk to soldiers without compromising their ability to perform their duties morally, then they have an ethical obligation to do so.[48] Strawser goes on to assert that this means we must be given some good reason *not* to use unmanned systems.[49] Here, he is trying to shift the burden of proof onto those who oppose the employment of unmanned systems.

Many will simply accept Strawser's account of why states should be considered obligated to employ unmanned systems, but there are a few criticisms to be made. The first is that it seems rather uncontroversial to suggest that if these technologies *exclusively* minimise harm to warfighters – that is, while imposing no additional 'costs' of another kind – it would be a morally good thing to use them. Most people would presumably agree that, given two equally good options for achieving some good goal, we are ethically required to select that which incurs the least lethal risk. The United States Air Force's operational risk management policy adopts this principle, stating that the most logical choice for any given task is that which meets task requirements while exposing personnel to the 'lowest acceptable risk'.[50] Strawser's account, which may well have been derived from that policy, simply formalises a common moral intuition and adds a long list of provisos and qualifications.

It is clearly also difficult to dispute that, in *some* cases, a military force would be obligated to use unmanned systems to reduce the risk of harm to their soldiers. The real worry is that things may not be as clear-cut or as perfectly fair as some premises of Strawser's argument seem to suppose, making it difficult to determine when this obligation exists. For instance, there will often be cases where we will be inclined to make delicate trade-offs between military capability and the protection of life. Imagine that the instigators of a morally sanctioned war were able to significantly reduce military casualties by using an unmanned system, but only by raising the probability of causing harm to civilians by some small degree. In such cases it simply is not obvious whether the use of unmanned weaponry would be permissible, let alone obligatory. Therefore, the problem is not so much with the normative consequent of Strawser's conditional claim, but rather the ease of determining whether the antecedent is true. In the military sphere with its many shades of grey, the antecedent cannot always be treated as, or reduced down to, a clear 'empirical question … that can be tested for its veracity',[51] as Strawser would have us believe. We must also ask ourselves whether Strawser is wise to shift the burden of proof onto those who oppose the use of unmanned systems. Do we really know enough about why we ought to use these systems to jump straight to the conclusion that we should use these systems and refrain only if some problem presents itself? In the absence of consensus, we should take a precautionary approach and stipulate that the burden of proof (to demonstrate that unmanned warfare is not unethical) should fall squarely on those that resort to such action.

Strawser grounds the obligations of military forces to employ unmanned systems solely on the capacity of these systems to minimise the risk of harm to soldiers. But is this the only, or most important, factor? Does it adequately express the full moral justification? He himself signals that there are other arguments that could support the case for the employment of unmanned systems. For instance, he notes that they are, on the whole, cheaper to produce and deploy than manned systems that accomplish similar missions (as outlined earlier). He recognises that one could argue that we are obligated to spend the least amount of shared resources possible on any given collective venture, including war, since these resources could be used for other worthy goals; and he says that we could develop a 'principle of unnecessary waste of scarce resources' to capture the normative appeal of such an approach.[52] However, Strawser sets these financial concerns aside early on, and fails to discuss any of the other benefits associated with the use of these systems – such as reducing the environmental impact or maximising military capability – instead choosing to focus on what he sees as the more important matter: the capacity of these unmanned systems to reduce the risk of harm faced by soldiers. He does so to the detriment of a fuller understanding of why it is important to use these systems.[53] That is because saving soldiers from harm is not, in fact, the most weighty reason, nor clearly the exclusive reason, behind the duty to employ unmanned systems; and to provide a more complete account of the duty to employ these systems we must give weight to all the benefits associated with their use and provide a fuller explanation of why states and their military forces ought to seek these benefits. In order to do so, we need to consider how these entities have come to power, what their main responsibilities are and how the benefits that unmanned systems provide aid in fulfilling these responsibilities.

The Nature of the Relationship between Citizen, State and Military Force

For us to fully comprehend the duty to employ unmanned systems, we first need to understand precisely in what the state's protective duties are grounded. We know that states currently have a monopoly on the waging of violence; however, we need to explain why states do, or indeed should, lay claim to the monopoly on the use of armed force before we can really understand the obligation to employ unmanned weapons. We must therefore return to the theoretical and ideological foundations of the state and consider the relationship between citizen, state and military force. This relationship can be quite complex, particularly in liberal Western democratic societies such as Australia, the United Kingdom and the United States, among the many others. In these societies, the relationship rests somewhat precariously on the rather imprecise notion of shared or common values and manifests itself in the form of a not particularly well-defined network of rights and duties.[54] The mechanism through which our relationship to other individuals and the state is regulated, and our responsibilities defined, has been termed the 'social contract'. This is essentially an implicit agreement that outlines how everyone is to gain mutual benefit by living together under the aegis of the state. Let us briefly summarise the social contractarian view.

For John Locke and Thomas Hobbes, both seventeenth to eighteenth century philosophers, the need for people to enter into an agreement or social contract with the state arises from what is termed the 'state of nature'. This is essentially an initial environment or situation in which individuals exist prior to the formation of a state. It is unclear whether these philosophers thought it to be an actual historical situation, a possible historical moment or a purely hypothetical scenario.[55] In any case, Locke says that the state of nature, the natural condition of humankind, is one of 'perfect freedom' in which people are free 'to order their actions, and dispose of their possessions and persons as they think fit, within the bounds of the law of nature, without asking leave, or depending upon the will of any other man'.[56] For Locke, this is because those in the state of nature are bound by God's 'law of nature', which obliges individuals not to harm others and to have regard for each other's lives, health, liberty and possessions.[57] However, Locke recognises that the peaceful state of nature can disintegrate into something resembling war when the conditions of the law of nature are violated. As there is no civil authority to enforce the law of nature and nobody to 'preserve the innocent and restrain offenders',[58] individuals must provide for their own security as best they can, thereby subjecting those without the means of protection to the will of those that do.

In his *Leviathan*, Hobbes presents a rather more frightening and quite grim picture of the state of nature.[59] For him, a number of 'natural' factors or conditions conspire against individuals in the state of nature, placing them in hostile competition with each other. First, there is an equality of need. Everyone has the same basic need for food, water, shelter and so on. This equality of need, in itself, would not be a problem if it were not for the second factor: the scarcity of these resources. There are simply not enough of these basic resources to fulfil everyone's needs. For a short while, a few might succeed in achieving power and control, taking whatever they want at the expense of all others, but – and this is the third natural condition – individuals in the state of nature are inherently equal in their powers. Some might be stronger or smarter than others, but every human being, according to Hobbes, has the same ability to attain their ends and an equal capacity to harm and kill, leaving all against all. A possible solution would be to rely on the altruistic kindness of others, but for Hobbes, this is not an option because of the fourth natural condition: people are fundamentally self-interested. He therefore finds the state of nature to be unbearable and life within it to be 'solitary, poor, nasty, brutish and short'.[60] In the state of nature he describes, life is reduced to a constant struggle for survival, since without a common power, those within it exist in a state of anarchy.[61]

Those in the state of nature are not, however, in a totally hopeless position. There is a way out of the thoroughly undesirable initial condition that is the state of nature. Both Locke and Hobbes suggest that self-interested but rational individuals in the state of nature would think it best to set up a mechanism to provide safety and security for all. Heavily influenced by the armed conflicts of the English Civil War, Hobbes wrote that the only way for those in the state of nature to protect themselves over the long-term is to erect a common power to defend themselves from outside invasion and from the injuries of one another. The existence of a state/common power in turn enables people go about nourishing themselves and enjoying the 'fruit[e]s of the Earth'.[62] To afford themselves this better life, they must construct a social contract. This contract involves the signatories agreeing to two things. Firstly, they must agree to establish a better society by collectively and reciprocally renouncing any other more primitive rights that they might have held against one another in the state of nature. Secondly, and perhaps most importantly, they must imbue some person or group of persons with the power and authority to enforce adherence to the contract. Essentially, to progress beyond the state of nature, they must agree to live under common rule and create a sovereign power to enforce the agreement.

The social contract is supposed to define and regulate the relationship between the people and this central power, that is, between individuals and the newly created sovereign. Under the contract, individuals give up their natural right to the use of armed force for their private protection and entrust their security and wellbeing to the sovereign.[63] The social contract therefore obliges the sovereign to try to attain security and justice for all, which, in the first instance, means ensuring that all those within its territory respect the rule of law. The sovereign will therefore monopolise the use of force internally, to encourage individuals to obey the covenants that are in place and respect the lives, liberties and property of others, threatening them with punishment if they do not.[64] On Hobbes's account of the social contract, the sovereign state is also conceived as having a monopoly over the means and use of force against *external* threats, whether these are other sovereigns or informal collectives. It is responsible for raising armies and decides when and how to wage war. According to all three accounts of Hobbes's political theory, the *Elements of Law*, *De Cive* and the already cited *Leviathan*, it is also in the sovereign's purview to command people to serve as instruments of these armies.[65]

Jean-Jacques Rousseau built on Hobbes's conception of the social contract,[66] but put forward a somewhat more moderate account that arguably leads to ideas with greater relevance for today's armed forces. His account was of a pact between equal citizens, rather than an agreement between people and a ruling sovereign that is largely detached from the population.[67] For Rousseau, the social contract should provide for the security and wellbeing of citizens, as per Hobbes's interpretation, but should allow individuals to retain their freedom and rights of self-determination. So, unlike Hobbes, for whom the sovereign is an external entity placed above society, Rousseau considers the people within a territory to be the sovereign. The state does not stand above society as some external coercive power but rather evolves from society itself. It is essentially the merger of all individual wills into a general will, which is more than, and different to, the mere aggregation of the individual wills.[68] Rousseau states that under the social contract, individuals lose their natural liberty and their unlimited right to whatever they attempt to get and succeed in getting, which is bounded only by the strength of the individual; what they gain is civil liberty and the proprietorship of all they possess, which is, instead of by the individual, limited by the general will.[69] This means that rather than politico-social decisions being made by a sovereign that is distinct and detached from people, citizens remain free to determine their own fate by transgressing laws of the larger community at their own cost.[70]

Rousseau's liberal democratic social contract has to be considered an agreement by which citizens of a nation come together to create a sovereign state entity, the sole function of which is to serve as an instrument of citizens' interests. It is therefore obliged to enforce law and order domestically. Furthermore, because the state has a wider obligation to protect the rights, lives and property of citizens, and because it is tragically true that some human societies feel the need to destroy other human societies, it must establish standing armed forces. For Rousseau, individuals may, at times, be required to share the burdens of citizenship by serving in the armed forces under the authority of the general will and in protection of the common life.[71] It is this sense that Rousseau's conception more accurately reflects the modern conditions under which soldiers enlist or are conscripted and the values to which they are expected to adhere.

It is important to pay careful attention to these social contract theories – among others from Socrates through to modern philosophers like John Rawls – because they form the most influential body of thought that addresses why states exist and what, exactly, they exist for.[72] Despite there being many different variants, each with its own important differences, all agree that the primary justification for the existence of states revolves around their role in providing safety and security for those living within their territory. For all intents and purposes, states monopolise the right to wage armed violence through claiming to be the *most capable* agent for successfully representing and defending the general interests of those they govern.[73] It is essentially an argument based upon ensuring citizens' safety. As Cécile Fabre writes to support her cosmopolitan account of state legitimacy: 'the state's right to wage war is one which it has precisely in so far as it is better than individuals at protecting their fundamental human rights through the use of lethal force'.[74] This protective role is why a state's use of force in defence of its citizens and their common life has positive moral standing in just war theory (to be discussed in the next chapter) and international law. This is not to say that all assertions of the collective right to defence stand unchallenged;[75] however, in what follows, let us presuppose the existence of this right and consider that it is derived from consideration of the interests of the people and exercised on their behalf.

The Military-State Contract: Grounding the Duty to Employ Unmanned Systems

As detailed in the previous section, social contract theory provides a compelling account of the existence of states and their military forces, and of how the social contract defines and regulates the relationship between individuals and the state. However, this contract does not extend explicitly to the military and can only begin to explain how the practical benefits and the prudential argument put forward in the first section, could turn into a moral contract. We must, therefore, posit the existence of a second contract to explain the duty to employ unmanned systems. This can usefully be construed as an implicit contract that, despite being somewhat obscure, is often used implicitly by states and military forces to justify how and why they invest their limited resources in particular (and often controversial) ways. So as to avoid confusion with both the social contract discussed above and the standard civil-military covenant, which is primarily concerned with interactions at the individual level (especially between states and their veterans), let us label this the 'military-state contract'.[76] The military-state contract regulates the relationship between the state and its military institutions as abstract entities.

As Peter Feaver has argued, in any liberal democracy, this relationship is one of delegation.[77] The state delegates or contracts out much of its protective function to the military and its armed forces, just as the citizenry entrusts many of its responsibilities to the state.[78] In assigning these critical protective duties to the armed forces, the state is essentially using it as an agent. Together, they share the responsibility of protecting the citizenry. To ensure that society is adequately safeguarded under this arrangement, the military-state contract assigns mutual obligations and duties to the involved parties. The state's primary duty under the contract is to provide adequate funding for the armed forces and to enact law, which enable them to go about fulfilling their protective function. In return, the military has an obligation to effectively and efficiently provide for the safety and security of the state and its people, while ensuring that it does not against the people's broader interests.[79] In other words, prudential action in the form of ensuring effectiveness and efficiency in military operations becomes a moral requirement. Also recall from the previous section, and from Fabre's claim in particular, that a state's claim to legitimacy is grounded on it being the most efficient and well-equipped entity to achieve security from internal and external violence and harm. If military institutions mismanage their relationship with the state and are neither effective, nor efficient, in protecting society against potential threats, it jeopardises or erodes the initial social contract between citizen and state. The state and its military institutions should expect to be distrusted under such circumstances and should not be surprised when punitive measures are invoked.[80]

Since the United States has been trying to shift toward a more measured use of violence as a result of the 'needless waste' of Vietnam,[81] one might think that there would be a wealth of literature on the contractual obligation to ensure military effectiveness and efficiency. However, a review of the relevant literature reveals that there is very little published work on the topic. While key historical works such as Sun Tzu's *Art of War*[82] and Carl von Clausewitz's *Principles of War*[83] and *On War*[84] deal with effectiveness and efficiency in some detail and lend some support to the notion of a military-state contract described above, they do so only in an indirect way. Indeed, they are more concerned with strategy, tactics and the general nature of war than with the military-state relationship itself. Even in the wide body of modern 'civil-military relations' theory, which seeks to explain the hierarchical interactions that take place between civil society and the military organisation or organisations that are charged with protecting it, the obligations of effectiveness and efficiency are under-analysed. To illustrate this point, Samuel Huntington's *The Soldier and the State*[85] – the standard reference point in civil-military relations theory – contains only a single passing reference to effectiveness and makes no reference at all to efficiency.[86] Although civil-military relations are extremely complex, the vast majority of civil-military relations literature has solely focused on how to control the armed forces and prevent them from asserting their will on those they are meant to protect.[87]

Thomas Bruneau recognises this fact in his most recent book, *Patriots for Profit.* There, he extends civil-military relations to include effectiveness and efficiency in order to explain why, for more than 30 years, Western military forces (most notably those of the United States and the United Kingdom)[88] have contracted out many of their functions rather than hiring government employees to perform them.[89] In particular, Bruneau's focus is on why private military contractors have taken on many of the combat roles and missions that were traditionally left for uniformed military personnel.[90] While Bruneau's subject matter is different, he argues that effectiveness (in terms of implementing roles) and efficiency (in the use of funds and personnel) are critical to understanding the changes that today's military forces are undergoing, technological or otherwise.[91] He also points out that America already has multiple robust institutions aimed at ensuring that effectiveness and efficiency is attained, but that a lack of general context (such as that provided by the contract advocated here) and interagency process is responsible for poor outcomes.[92] Therefore, Bruneau's general argument seems to lend further support to the military-state contract formulated earlier as well as to any account that attempts to establish or formalise effectiveness and efficiency as moral imperatives in the contractual relationship between military and state. Prior to outlining some key obligations under the military-state contract, it is necessary to briefly elucidate Bruneau's definition of effectiveness and efficiency, because in common parlance the terms are often used inconsistently and interchangeably.

Quite rightly, Bruneau states that an armed force's effectiveness is based on its ability to 'implement the roles and missions assigned to it by democratically elected civilians', while its efficiency is a matter of achieving those 'roles and missions … at the least possible cost in lives and resources'.[93] From these definitions we can derive a few key criteria that a military force must satisfy to be classified as meeting its requirements under the proposed military-state contract. For the moment, we will consider various obligations to ensure efficiency. As per Strawser's account, a military force must seek to minimise the risk of harm to its military personnel. This is because soldiers' lives are of *instrumental* value to the armed forces and the state, and by proxy, to everyone in society. In playing a role in securing or protecting the state, or otherwise advancing the state's goals, they are performing an important public function. For their period of service, soldiers are instruments of the state or, more accurately, its military.[94]

In this respect, soldiers can be seen as precious resources that should not be wasted, primarily because they are hard to procure and are necessary to protect the state and secure the common life. Hence, part of the value of reducing the risk of harm to soldiers comes from preserving their ability to partake in future military operations, whether in some capacity on the ground within hostile territory or on home soil in front of a monitor. This is one reason why serving and retired military personnel typically receive free health care that is equal to or better than that received by ordinary citizens: it is, in part, the state's insurance policy for its future soldiers.[95] At the same time, soldiers are more than mere assets or replaceable cogs in the military machine and we have to introduce their intrinsic moral value into the calculation.[96] We have to remember that armed forces serve states, that states serve citizens, and that every soldier is essentially a 'citizen in uniform'.[97] In Western liberal democracies, citizens are not usually forced to serve, but do so out of a sense of individual obligation induced by the shared values of the society in which they live.[98] As Martin Cook puts it, they render their service as one way of sharing the burdens of citizenship and protecting the 'common life' that they share with fellow citizens, perhaps explaining why soldiering is often described as a 'higher calling'.[99] Thus to deliberately or unnecessarily risk a soldier's life would not only be to squander a precious resource, but also to place a disproportionate contributory burden on the citizen, which is yet another reason why a legitimate military force should constantly be striving to minimise risks.[100]

Under the military-state contract, military institutions must also ensure that they use their financial resources as wisely as possible. The leaders of the military ought to be recognised as the trustees and fiduciaries of the common assets put aside to ensure the safety of the community. Some military forces, particularly those of wealthy Western states, often do not seem to concern themselves with the financial cost of their operations. As Andrew Bacevich writes, in some outfits, there would appear to be a 'general rule' that 'when the issue is a military one, money is no object'.[101] He goes on to note that military forces tend to have a willingness to undertake bold initiatives and an unusual tolerance for cost overruns and general financial inefficiency.[102] The United States military's purchase of $300 hammers, $600 toilet seats[103] and its more recent multi-million dollar purchase of 'ballistic undergarments'[104] seem to confirm this assertion, at least in relation to past years. More recently, this liberal spending has been tempered by political pressure. Politicians are now constantly reminding their military forces of the financial costs of war. In a speech given to the United States Military Academy at West Point, President Barack Obama stated that (referring to the trillion dollar price tag of the 'War on Terrorism'),[105] 'in the wake of an economic crisis, too many of our neighbors and friends are out of work and struggle to pay the bills … so we can't afford to ignore the price of these wars'.[106]

In a similar respect, the contractual account advanced here reminds us that while a state cannot afford to underfund a military responsible for maintaining a standing army at an appropriate level of force readiness, which entails procuring the tools and resources necessary for protecting the state and its citizenry, considerations of financial efficiency cannot be put aside. To sustain the services provided directly or indirectly by the state, citizens, for their part in the social contract, have an obligation to pay taxes. As outlined earlier, the state allocates a portion of these public funds to its military institutions for the purpose of providing security from violence. The military-state contract dictates that the portion of taxpayer funds allocated to the provision of security be used judiciously. When financial resources are limited, the taxpayer funds can, if not used by the military, be put toward some other life-saving effort or social good, protecting thus the interests of the state and its citizens.[107] The military-state contract therefore supports the idea of an obligation on the part of armed forces to ensure they use these funds to pursue only worthy goals which they are specifically obliged to follow using the most reasonable and cost effective means.

Importantly, it also follows that a military force properly fulfilling its duties ought to respect the natural environment. While protection of the environment is obviously not the primary responsibility of the armed forces, unless it is threatened to be destroyed by another state as a means of warfare, the military-state contract requires military forces to make every effort to reduce or mitigate their impact on the environment when operational circumstances permit. This is because the military is an agent of the state, which is, above all else, responsible for ensuring that the basic needs of its citizens are met and that their general interests are advanced. And, while humanity's relationship with the environment is complex, it can be advanced that we should endeavour to preserve the environment because of its instrumental value in our welfare.[108] Apart from clean air, water and food, society and its military-industrial complex would not have been be able to flourish and develop as successfully as they have over the past few centuries without the extraction, collection and refinement of many natural resources, particularly crude oil. Indeed, it may be argued that our relationship to oil remains a volatile, yet necessary one. The natural environment with its abundance of resources and possible intrinsic value (which many people think it has) is therefore not only central to our economic, social and personal wellbeing, but is also of fundamental moral significance under the social contract. As such, there are strong reasons to think that the environment should be protected by the state and its armed forces. This applies wherever they may operate. A military force has the potential to damage another state's environment so greatly that this harm would have repercussions on a global scale and so this would end up harming the very citizens this military force is meant to protect. For example, nuclear weapons can deposit radioactive particles into the atmosphere that can then fall anywhere around the globe.[109] Also, consider the large volumes of fossil fuels burnt during global military operations. They are thought to be depleting the ozone layer and contributing to global warming which, while impacting some states more dramatically than others, has consequences for all.[110] While the matter of 'future harms' is one area of substantial controversy, with the social contract requiring modification to account for future citizens, states must carefully consider their potential role in harming its future generations by causing irreparable damage to the environment.

The argument above shows that a state's military force therefore has a number of *prima facie* obligations under the military-state contract. It must: minimise the risk of harm to its military personnel; use those financial resources provided to it by the state wisely and judiciously; and minimise its impact on the shared natural environment. However, all of these are obligations regarding efficiency and, while an efficient application of force often yields the best operational outcome, a military force can be efficient without being effective.[111] For that reason, a military force's overriding obligation should always be to ensure that it is capable of fulfilling its primary organisational and operational purpose, which is to serve society, through the state, fighting the conflicts that the state endorses.[112] In other words, the ultimate determinant of whether a particular military force is meeting its obligations under the contract that governs its subservient relationship to the state and its people is whether it is capable of projecting power when and where necessary, or more precisely, fighting morally sanctioned wars and winning them. It is difficult to measure the extent to which a military force is capable of this, but it will often depend upon the weapons available to it. It can therefore be argued that to ensure the best possible chance of success against a variety of potential enemies, and across different operating environments, military forces have an obligation to continuously seek to design or embrace advantage-conferring weapons technologies to use when conducting their core business of defence. John Brennan, counterterrorism adviser to Barrack Obama, hinted at this during the first public acknowledgement of the US drone programme. He said that 'people expect us to use advanced technologies' in preventing attacks and saving lives.[113, 114]

It is from these obligations generated under the military-state contract that the duty to employ unmanned systems can be derived. That is, when you consider that a military force's sole duty is to provide effective and efficient protection and security, it becomes clearer why there exists a *prima facie* moral duty to employ unmanned systems. It is not so much about the intrinsic moral worth of soldiers' lives, as Strawser at times suggests, but more about ensuring that a military force can meet its obligations to the state and its many citizens, which have individual and collective moral worth. All of this is evident from a proper understanding of the military-state contract, which, with its focus on effectiveness and efficiency, establishes why the proposed benefits associated with the use of unmanned systems are of moral importance. As the beginning of this chapter outlined, these systems promise to reduce the human, financial and environmental costs of war, while ensuring that those armed forces that utilise such resources (human, financial and environmental) are more able to meet present and future challenges arising in relation to the defence of the citizens of a state and their common life. These are all benefits that a military force ought to seek with the use of any technology, and which ground the *prima facie* duty to employ unmanned systems.

Conclusion

The aim here has been to explore the moral reasoning behind the use of unmanned systems before moving on to consider the constraints that should apply to their use. The first section outlined some of the operational benefits that are thought to accompany the use of these systems. Indeed, when making a moral assessment of any technology, we must understand what it is that we stand to gain from its use. It was said that these systems hold potential to reduce the human, financial and environmental costs of warfare while enhancing warfighting capability. The second section looked at the relation between these benefits and the ethical justification for the use of unmanned systems. We began by considering Strawser's account of the obligation to employ these systems. His account holds that a military force has an ethical obligation to employ unmanned systems when the use of such systems minimises the risk of harm to its soldiers without incurring any relevant loss of capability. However, it was demonstrated that his account of the obligation to employ these systems ignores some of the complexities of using technology in the military sphere and only gives a partial account of why it is important to try to employ advantage-conferring technologies. To provide a more comprehensive account of why the use of unmanned systems is morally obligatory, the third section examined the contractual roots of the state's protective duties. Following from this, the fourth and final section argued that there exists a contract between state and military force, and that it is this 'military-state contract' which establishes prudential actions as a moral requirement for a legitimate military force, all of which fall under an overarching duty to provide the most effective and efficient protection and security possible to the state and its citizens. It was argued that it is this duty, coupled with the correlating benefits that unmanned systems provide, which grounds the case for their employment. It is this duty which must be kept in the foreground as we discuss the associated problems and one that we must remember is subject to strict conditions, as outlined in the following chapter.

Notes

1 For instance, see United States Department of Defense, 'Unmanned Systems Integrated Roadmap FY2011-2036' (Washington, DC: Department of Defense, 2011), 3; United States Department of Defense, 'Unmanned Systems Integrated Roadmap FY2013-2038' (Washington, DC: Department of Defense, 2013), v.

2 United States National Research Council, *Star 21: Strategic Technologies for the Army of the Twenty-First Century* (Washington, DC: National Academy Press, 1993), 27.

3 Susi V. Vassallo and Lewis S. Nelon, 'Thermoregulatory Principles', in *Goldfrank's Toxicologic Emergencies*, ed. Martin J. Wonsiewicz, Karen G. Edmonson, and Peter J. Boyle (New York: McGraw Hill, 2006), 265.

4 Lee Davis, *Natural Disasters* (New York: Facts on File, Inc., 2008), 7.

5 Christopher Coker, *The Warrior Ethos: Military Culture and the War on Terror* (New York: Routledge, 2007).

6 For examples of the effects of other toxic weapons, see Institute of Medicine, *Gulf War and Health: Depleted Uranium, Sarin, Pyridostigmine* (Washington, DC: National Academy Press, 2000).

7 James A. Tyner, *Military Legacies: A World Made by War* (New York: Routledge, 2010), 151.

8 Krishnan, *Killer Robots: Legality and Ethicality of Autonomous Weapons*, 123.

9 John Weckert, 'Risks and Scientific Responsibilities in Nanotechnology', in *Handbook of Risk Theory: Epistemology, Decision Theory, Ethics, and Social Implications of Risk*, ed. Sabine Roeser, et al. (Dordrecht: Springer, 2011), 161–2.

10 Paul D. Walker, *Battle Fatigue: Understanding PTSD and Finding a Cure* (Bloomington: iUniverse, 2008). 73.

11 Coker, *The Warrior Ethos: Military Culture and the War on Terror*, 100.

12 Lisa Daniel, 'Military Leads Mental Health Care Transformation', *American Forces Press Service*, 24 February 2010, 346.

13 Robert S. Rush, *Enlisted Soldier's Guide* (Mechanicsburg: Stackpole Books, 2006), 138.

14 Ethan Bronner, 'Israel Plans New Rules on Exchange of Prisoners', *The New York Times* (2012), http://www.nytimes.com/2012/01/06/world/middleeast/after-shalit-israel-changing-prisoner-exchange-rules.html.

15 Markus D. Dubber, *Victims in the War on Crime: The Use and Abuse of Victims' Rights* (New York: New York University Press, 2002), 284.

16 Andrew Hoskins, *Televising War: From Vietnam to Iraq* (London: Continuum International Publishing Group, 2004), 67.

17 Bronner, 'Israel Plans New Rules on Exchange of Prisoners'.

18 United Kingdom Ministry of Defence, 'Joint Doctrine Note 2/11: The UK Approach to Unmanned Aircraft Systems', 3–7.

19 United States National Research Council, *Autonomous Vehicles in Support of Naval Operations*, 97.

20 The number of personnel required to support the operation of the aircraft would be comparable between manned and unmanned systems. Both require refuelling, maintenance and a large number of data analysts.

21 Rui Goncalves et al., 'Authority Sharing in Mixed Initiative Control of Multiple Uninhabited Aerial Vehicles' (paper presented at the Engineering Psychology and Cognitive Ergonomics, Orlando, Florida, 2011), 535.

22 The cost of putting a single soldier on the battlefield is approximately US$1,000,000.

23 United States National Research Council, *Autonomous Vehicles in Support of Naval Operations*, 97.

24 Ibid.

25 Steven Graham, 'The New Military Urbanism', in *The New Blackwell Companion to the City*, ed. Gary Bridge and Sophie Watson (Malden: Wiley-Blackwell, 2011), 131.

26 D. Belin and Gary Chapman, *Computers in Battle: Will They Work?* (New York: Harcourt Brace Jovanovich Publishers, 1987), 76.

27 Bill Yenne, *Attack of the Drones: A History of Unmanned Aerial Combat* (Minneapolis: Zenith Press, 2004), 12.

28 Medea Benjamin, *Drone Warfare: Killing by Remote Control* (New York: OR Books, 2012), 20.

29 N. Turse, *The Complex: How the Military Invades Our Everyday Lives* (London: Faber & Faber, 2008), 41.

30 Geoff Martin and Erin Steuter, *Pop Culture Goes to War: Enlisting and Resisting Militarism in the War on Terror* (Plymouth: Lexington Books, 2010), 28.

31 W. G. R. Vieira et al., 'Structural Dynamics and Renewable Energy' (paper presented at the 28th IMAC, A Conference on Structural Dynamics, 2010), 248.

32 David Hay-Edie, 'The Military's Impact on the Environment: A Neglected Aspect of the Sustainable Development Debate' (Geneva: International Peace Bureau, 2002), 4.

33 Associated Press, 'US Military Damaged Ancient Babylon Site', *Sydney Morning Herald*, http://www.smh.com.au/world/us-military-damaged-ancient-babylon-site-20090710-deuh.html.

34 Arkin, *Governing Lethal Behavior in Autonomous Robots*, 29–30.

35 T. Weiner, 'GI Robot Rolls toward the Battlefield', *Herald Tribune*, 17 February 2005.

36 Christopher Coker, *Waging War without Warriors? The Changing Culture of Military Conflict* (Boulder, CO: Lynne Rienner Publishers, 2002), 59.

37 Women are often disqualified from service in combat positions because of their alleged: lack of physical strength; feminine hygiene and health requirements; effect on team cohesion and so on. As the distance from the battlefield increases, these factors become less significant and/or easier to accommodate. For instance, some uninhabited aerial vehicle operators work shifts in what is not all that different to a normal office environment. There is no reason why women and the disabled cannot be equally as effective in these

sort of roles. Also note that Australia has only recently given women access to all combat roles, see Mark Dodd, 'Combat Roles Offered to Women', *The Australian*, 12 April 2011.

38 Bradley Jay Strawser, 'Moral Predators: The Duty to Employ Uninhabited Aerial Vehicles', *Journal of Military Ethics* 9, no. 4 (2010). Strawser talks exclusively about unmanned aerial vehicles; however, his argument applies equally well to the broader category of unmanned systems. His argument is generalised here.

39 The giving of an order is important, as the question of whether it is permissible for an individual to take on unnecessary risk in the achievement of some good goal is a different matter (although any answer to this question would have to take into consideration that the individual in the above example is rendering a service to the state).

40 Strawser, 'Moral Predators: The Duty to Employ Uninhabited Aerial Vehicles', 344.

41 Ibid., 345.

42 If it were to aid in the achievement of the goal the matter would be complicated as this would offset the duty to avoid incurring lethal risk.

43 Strawser, 'Moral Predators: The Duty to Employ Uninhabited Aerial Vehicles', 345.

44 Ibid., 346.

45 Ibid., 347–8.

46 Ibid.

47 Some might take issue with the analogy used to support his strong claim for the use of these systems. Bomb squads use unmanned systems to disarm or destroy explosive ordinance, in order to *prevent* the loss of life, whereas a military force using unmanned systems will, in many cases, be doing so in an attempt to *take* life, even if it is in self-defence. The complication therefore seems to result from moving from the civilian case to the military case. However, as Strawser rightly suggests, the point of disanalogy between the two cases is insignificant, because while military forces may use these systems to take life, this is done as part of (what should be) a morally justified conflict. The main point is that they contribute to some good, and in doing so, remove some of the risk the agent would otherwise face in carrying out the action, as is the case with those remote systems that are used by civilians to disarm explosives. See ibid., 348.

48 Ibid.

49 Ibid., 349.

50 Francis C. Gideon, *Air Force Instruction 90–901* (Washington, DC: United States Department of Defense, 2000).

51 Strawser, 'Moral Predators: The Duty to Employ Uninhabited Aerial Vehicles', 347.

52 Ibid., 344.

53 It is my understanding that Strawser is in the process of developing his views to resolve this problem.

54 Michael Bryce, 'The Unique Nature of Military Service', in *Defence Force Welfare Association Forum* (Canberra: Defence Force Welfare Association, 2009).

55 George Klosko, 'Contemporary Anglo-American Political Philosophy', in *The Oxford Handbook of the History of Political Philosophy*, ed. George Klosko (Oxford: Oxford University Press, 2011), 459.

56 John Locke, *The Two Treatises of Government* (Indianapolis: Hackett Publishing Company, 1980), 8.

57 Ibid.

58 Ibid., 9. See remainder of Chapter 2 for Locke's discussion of the state of nature.

59 Thomas Hobbes, *Leviathan* (Sioux Falls: Nuvision Publishers, 2004).

60 Ibid., 74.

61 Ibid., see Chapter 13 for more detail.

62 Ibid., 99.

63 E. Krahmann, *State, Citizens and the Privatization of Security* (Cambridge: Cambridge University Press, 2010), 23.

64 Hobbes, *Leviathan*: 84.

65 D. Baumgold, 'Subjects and Soldiers: Hobbes on Military Service', *History of Political Thought* 4, no. 1 (1983): 43.

66 Jean-Jacques Rousseau, *On the Social Contract*, trans. G. D. H. Cole (Mineola: Courier Dover Publications, 2003).

67 Ibid., 12.

68 Ibid., 17.

69 Ibid., 12.

70 Ibid., 67.

71 Krahmann, *State, Citizens and the Privatization of Security*, 26.

72 Deane-Peter Baker, *Just Warriors Inc.: The Ethics of Privatised Force* (New York: Continuum, 2011), 67.

73 Patricia Owens, 'Distinctions, Distinctions: "Public" and "Private" Force?', *International Affairs* 84, no. 5 (2008): 980–81.

74 Cécile Fabre, 'Cosmopolitanism, Just War Theory and Legitimate Authority', *Journal of International Affairs* 84, no. 5 (2008): 972.

75 For a comprehensive outline of some of the issues, see David Rodin, *War and Self-Defense* (Oxford: Clarendon, 2002), 66–84; For a response, see Baker, *Just Warriors Inc.: The Ethics of Privatised Force.*

76 It is important to note that the scope here is fairly limited and that this is not an attempt to formalise all civil-military relations into this contract.

77 Peter D. Feaver, *Armed Servants: Agency, Oversight, and Civil-Military Relations* (Cambridge: Harvard University Press, 2003), 57.

78 Deane-Peter Baker, 'To Whom Does a Private Military Commander Owe Allegiance?', in *New Wars and New Soldiers: Military Ethics in the Contemporary World*, ed. Paolo Tripodi and Jessica Wolfendale (Farnham: Ashgate, 2011), 186.

79 It must be acknowledged that state security may be different to people's security and that serving 'the people's broader interests' could justify nearly anything, including violent coups and so on. However, this claim is made in relation to representative democracies and, in referring to 'broader interests', concerns the basic necessities for life, such as a habitable environment.

80 While there are some key differences, there is a basic parallel here with the role of corporations. On many accounts, they exist primarily to create profits for their shareholders. If a corporation fails to deliver profits, its shareholders can legitimately overthrow those responsible for managing it, on grounds that they have not fulfilled the corporation's core duty.

81 Charles Dunlap, 'Organizational Change and the New Technologies of War', in *Joint Services Conference on Professional Ethics* (Washington, DC, 1998), 11.

82 Sun Tzu, *The Art of War* (London: Wordsworth Editions, 1990).

83 Carl von Clausewitz, *Principles of War*, trans. Hans W. Gatzke (Harrisburg: The Military Service Publishing Company, 1942).

84 Carl von Clausewitz, *On War*, trans. O.J. Matthijis Jolles (New York: The Modern Library, 1943).

85 Samuel P. Huntington, *The Soldier and the State: The Theory and Politics of Civil-Military Relations* (Cambridge: Belknap Press, 1981).

86 Ibid., 3.

87 Peter D. Feaver, 'Civil-Military Relations', *Annual Review of Political Science* 2, no. 1 (1999): 211; Dale R. Herspring, *The Pentagon and the Presidency: Civil-Military Relations from FDR to George W. Bush* (Lawrence: University of Kansas Press, 2005). xii.

88 Interestingly, the UK-based Serco Group is the largest government services contractor in the world, performing functions such as cleaning and cooking to firefighting and live range operations for military forces around the globe, including those of Australia.

89 Thomas C. Bruneau, *Patriots for Profit: Contractors and the Military in the U.S. National Security* (Stanford: Stanford University Press, 2011).

90 Ibid., 3, also see Chapter 5 for the reasons and legal bases.

91 Ibid., Chapter 3.

92 Ibid., 73–74.

93 Thomas C. Bruneau, 'Civil-Military Relations in Latin America: The Hedgehog and the Fox Revisited', *Revista Fuerzas Armadas y Sociedad* 19, no. 1 (2005): 123.

94 Huntington, *The Soldier and the State: The Theory and Politics of Civil-Military Relations*, 73.

95 While health care may well be provided as a benefit and repayment for providing military service, in an age where some countries permit their military personnel to retire after only 20 years' service (meaning that some retirees could be as young as 38 years of age if they joined the military early in life), we would be naïve to think that there are no other motives at play.

96 David J. Garren, 'Soldiers, Slaves and the Liberal State', *Philosophy & Public Policy Quarterly* 27, no. 1/2 (2007): 9.

97 After World War II, Germany used this label to allay militaristic fears and to encourage the world public to trust its soldiers. See Harald Müller, 'No Need to Fear United Germany', *Bulletin of the Atomic Scientists* 46, no. 3 (1990): 15; Harald Müller et al., 'Demokratie, Streitkräfte Und Militärische Einsätze: Der "Zweite Gesellschaftsvertrag"

Steht Auf Dem Spiel' (Frankfurt: Hessische Stiftung Friedens- und Konfliktforschung, 2010).

98 Bryce, 'The Unique Nature of Military Service', 3. The obvious exception is conscription, but it could be argued that even conscripted soldiers share some sense of the obligation induced by shared values and that it is simply overridden by other self-interested concerns.

99 Martin Cook, *The Moral Warrior: Ethics and Service in the U.S. Military* (Albany: State University of New York Press, 2004), 51.

100 For further reflection on this topic, see Ned Dobos, 'Are Our Soldiers Assets or Workers?', *Sydney Morning Herald*, 4 June 2012.

101 Andrew J. Bacevich, *The New American Militarism: How Americans Are Seduced by War* (Oxford: Oxford University Press, 2005), 217.

102 Ibid.

103 Stephen Moore, 'The Federal Budget 10 Years Later: The Triumph of Big Government', in *The Republican Revolution 10 Years Later: Small Government or Business as Usual?*, ed. Chris R. Edwards and John C. Samples (Washington, DC: CATO Institute, 2005), 68.

104 Nic Halverson, "Marine Corps Orders Bulletproof Underwear," *Discovery News* (2011), http://news.discovery.com/tech/marine-corps-orders-bulletproof-underwear-110418.html.

105 Many suggest that the figure is actually much higher, see Joseph Stiglitz and Linda Bilmes, *The Three Trillion Dollar War* (New York: W. W. Norton & Company, 2008).

106 Office of the Press Secretary, 'Remarks by the President in Address to the Nation on the Way Forward in Afghanistan and Pakistan', The White House, http://www.whitehouse.gov/the-press-office/remarks-president-address-nation-way-forward-afghanistan-and-pakistan.

107 Andrew Samuel and John Weir, *Introduction to Engineering Design: Modelling, Synthesis and Problem Solving Strategies* (Burlington: Elsevier Butterworth-Heinemann, 2005), 347.

108 Note that this is not necessarily to deny that we should also preserve the environment on the basis of its intrinsic or non-instrumental value. See later in the paragraph.

109 National Research Council (US). Committee on the Effects of Nuclear Earth-Penetrator and Other Weapons, *Effects of Nulcear Earth-Penetrator and Other Weapons* (Washington, DC: National Academies Press, 2005).

110 Stephen Anderson, K. Madhava Sarma and Kristen Taddonio, *Technology Transfer for the Ozone Layer: Lessons for Climate Change* (London: Earthscan, 2007). Chapter 5, 'Military and Space Agency Leadership to Protect the Ozone Layer', might be of particular interest; Krishnan, *Killer Robots: Legality and Ethicality of Autonomous Weapons*: 121.

111 Thomas C. Bruneau and Florina C. Matei, 'Towards a New Conceptualization of Democratization and Civil-Military Relations', *Democratization* 15, no. 5 (2008): 923.

112 Don Snider, John Nagl and Tony Pfaff, *Army Professionalism, the Military Ethic, and Officership in the 21st Century* (Carlisle: The Strategic Studies Institute, 1999), 21.

113 John O. Brennan, 'The Efficacy and Ethics of U.S. Counterterrorism Strategy', *International Security Studies* (Woodrow Wilson International Center for Scholars, 2012).

114 It is also important to note that the obligation can be overextended (that is, taken too far). There is a point at which developing new weapons does not confer an advantage. This might be for strategic reasons: if State X develops a new weapon, State Y might follow suit and harness it in more disturbing ways, meaning that it would have been better not to develop it in the first instance. It might also be for prudential reasons related to operational yield. If an existing nuclear weapon yields only slightly better results than the one on the blueprints, the case for developing it is less convincing. Thus it seems possible to build a case for some sort of self-imposed arms control from the obligations to effectiveness and efficiency.

Chapter 4
Just Unmanned Warfare: Old Rules for New Wars?

In the last chapter, it was argued that states and their military institutions have a duty to employ unmanned systems as a result of obligations owed to their citizens under the military-state contract. This chapter explores the moral restraint that should be exercised in the use of these systems under this obligation. The first section introduces just war theory[1] as the moral framework of choice for examining the problems associated with the use of unmanned systems. It places the theory in context and, by way of reference to philosophers such as Immanuel Kant, Francisco Suárez and John Rawls, argues that it can fruitfully be grounded in contractual theory. The second and third sections acquaint the reader with the traditional 'just war duo'. That is, the two sets of principles that just war theory has developed into over time: *jus ad bellum* and *jus in bello*. The former applies to the decision *to go to* war and includes the principles of just cause, right authority, right intent, likelihood of success, proportionality (in relation to the significance of the harm) and last resort.[2] The latter is supposed to govern conduct *in* war and includes the principles of proportionality (in relation to the amount of force needed to achieve operational objectives) and discrimination.[3] The fourth and final section will take seriously the question of just war theory's contemporary relevance and this theory will be defended as the most useful tool for assessing the moral justifiability of the use of unmanned systems. This assessment will briefly consider Jeff McMahan's revisionist challenge and, in particular, his claims against the logical independence of the two traditional branches of classical just war theory.

Just War Theory: Context and Groundings

While a number of fundamental moral concepts could be employed in making an ethical assessment of unmanned systems, one conceptual apparatus is particularly well suited to the task and will be pressed into service in this analysis of unmanned systems. That apparatus is the theoretical framework of *just war*, which has strong Greco-Roman and Judeo-Christian roots and continues to pervade modern Western thought, albeit less permissive about the goods that can permissibly be sought than its original incarnation. As Duane Cady explains, the theory of just war maintains that war is both morally justifiable in principle and often morally justified in fact.[4] In other words, the basic notion that underlies the just war framework is that war is *sometimes* morally acceptable. Of course, if we grant this, we need some account of *when* war is permissible and when it is not, and of *what* is permissible in war and what is not. The principles of just war theory provide that account and are supposed to distinguish just from unjust wars and just from unjust means of waging war. In essence, their function, as Cady puts it, is to restrain both 'an overeager and overactive interest'[5] in *participation in war* and in the *means of war* once war has commenced. When placed in the broader conceptual landscape, just war theory offers us middle ground between two major opposing positions concerning war: pacifism and realism. A full discussion of these theoretical commitments is well beyond the remit of this chapter, but before going on to discuss the philosophical underpinnings of just war theory, it must be briefly explained why an analysis of the implications of utilising unmanned systems is unlikely to come from either of these competing theories.

According to Andrew Fiala, pacifism, in its standard and absolute form, involves a maximal and universal rejection of armed warfare.[6] For a pacifist of this variety, war is wrong, in principle and in fact. Pacifist moral opposition to warfare can spring from a range of possible rationales and foundations. They can be of a deeply religious nature; a consequentialist nature which claims that engaging in warfare is always going to be contrary to the end goal of achieving the best possible consequences; or a deontological nature, which claims that warfare is always and everywhere immoral as it infringes upon inviolable rights, whether of humans or the environment. The unconditional nature of absolute pacifism is problematic. It is true that an indomitable will to avoid violence can, in some instances, be just as useful in effecting positive change as the exertion of physical force (as demonstrated by figures such as Gandhi and Martin Luther King). It is also true that nonviolent options ought to be used whenever practicable for reasons of efficiency and harm minimisation. And, as Douglas Lackey argues, the view that most wars throughout history have done more harm than good is one that many would be willing to endorse.[7] However, it must be acknowledged that whatever may have been the case in the past, the possibility exists that an abhorrent aggressor will threaten the welfare and peace of innocent persons, and that in such cases, a war of defence needs to potentially justifiable. Its idealistic nature explains why pacifist arguments will put be aside for the sake of this analysis of unmanned systems: if granted, the analysis would never get off the ground, for the introduction of new weapons makes no difference to a committed pacifist who will maintain that war is wrong. This is not an argument against pacifism, but merely to say that it has little to offer any assessment of unmanned systems.

Much the same can be said of war realism. While pacifist theory claims that war is always immoral, realist theory claims that war is non-moral.[8] The main claim behind this theory, as noted by Grotius, is that there is no significant moral relationship between the states in an international society of the kind that exists between the people in a given society.[9] That is, the states belong in something resembling a state of nature in relation to each other while people do not. When wars inevitably occur, states act purely and appropriately out of self-interest. This does not mean that all forms of restraint are abandoned, but rather that restraint is only exercised when it is in the interests of the state.[10] However, contrary to realism and its dwindling number of academic advocates, one can find many examples in support of the idea that war and the use of weapons need to be restrained beyond the limits imposed by enlightened self-interest. Vietnam, with its widespread destruction and enduring human toll, comes to mind as a prime example. Moreover, Cady finds evidence against realism in the fact that theory does not coincide with practice and that the vast majority of people have long held that war *is* an appropriate object of moral judgement.[11] To extend his argument, Cady points out that one of the more devastating indictments of realist thought is that even soldiers and statesmen (as opposed to civilians) find their decisions subject to moral scrutiny. One might question whether this is so, but this appears to be confirmed by the fact that moral language now pervades global politics, as presented by both the media and politicians. In other words, by paying 'lip service'[12] to the importance of the rules of just war, states and their representatives seem to acknowledge that war is a morally governed activity, even though they often act in ways that most would consider to be in violation of just war rules. But, even if one is swayed by the realist account, it must be recognised that it too has little to add to the unmanned systems debate, because they are mere tools to be exploited in pursuing the interests of the state.

Having indicated why neither pacifists nor realists will find their commitments supported in this analysis of unmanned systems, we can now examine the philosophical underpinnings of just war theory, which accepts presuppositions made in the introduction of this book: namely that war *is* a moral question and that war *can* be justifiable under some circumstances. Just war theory has rich philosophical roots and can be grounded in a variety of ethical frameworks.[13] Theorists such as Richard Brandt and R.M. Hare base their defence of just war theory primarily on utilitarian principles.[14] That is, at some level, they determine the morality of war by reference to the maximisation and optimisation of desirable outcomes. However, this consequentialist grounding of just war theory can be quite problematic. As Tony Coady points out, there are some moral imperatives, such as the prohibition on using unmanned systems to knowingly kill noncombatants, which are not principally justifiable on the basis of their beneficial outcomes.[15] Also, under a purely utilitarian theory, there would seem to be no universally stipulable upper limit as to what can be done to achieve the end of a given war, because avoidance of dire consequences can be used to justify any range of dire actions. However, a state surely cannot do *anything* to win a war.[16] This is not to say that consequences are irrelevant, but simply that we should not make them our sole consideration. There are other considerations beyond consequences that are of moral relevance. To that end, there are plenty of non-consequentialist philosophical versions of just war theory. Some writers have supported the theory by appealing to particular religious traditions.[17] The United States Catholic Bishops have, for instance, put forward their own defence of just war theory.[18] Other writers have appealed to the logic of *prima facie* duties and 'mid-level intuitive' principles such as non-maleficence and beneficence to generate just war principles,[19] while prominent writers such as Walzer primarily employ 'rights talk' in developing just war principles.[20, 21]

Most interesting for our purposes here is that just war theory can also be conceptually grounded in contract theory, capturing the notion of the military-state contract discussed in the previous chapter. The essence of the most basic form of the contractual theory of war can be stated fairly simply: it rejects more mainstream models of war in favour of one that views warfare as a contract or agreement between two or more independent states to settle their quarrel(s) by force of arms.[22] That is, it presents a model of war that in some respects resembles a duel writ large. Historically, contractual accounts of war have received little attention and have appeared in the legal and philosophical literature in what can, at best, be described as a somewhat disappointing piecemeal fashion. The idea of a 'war contract' seems to have as one of its earlier sources, the work of sixteenth-century philosopher and Spanish Jesuit theologian and just war theorist, Francisco Suárez.[23] In one passage in his treatise *On War*, he considers the case of two disgruntled princes who, unable to settle their dispute with each other by means other than war, 'join in a pact to the effect that the victorious party shall acquire the property of the vanquished'.[24] In the late seventeenth century, German jurist and political philosopher Samuel von Pufendorf also wrote about the war contract. As he explains it: 'the belligerents at the outset [form] an agreement to rest their case with the fortune of battle'.[25] His view was that 'practically all formal wars appear to suppose an agreement that he upon whose side the fortune of war has rested can impose his entire will upon the conquered'.[26] As will later be shown, this is not a view maintained in traditional just war theory. For a more moderate view, we have to go to Immanuel Kant, who also likened war to a contract.

In *The Metaphysics of Morals*, Kant claims that 'if one wishes to find any rights in wartime, one must assume the existence of something analogous to a contract; in other words, one must assume that the other party has accepted the declaration of war and that both parties therefore wish to prosecute their rights in this manner'.[27] However, as Peter Temes writes, to understand Kant's views on war, we must begin with his overall moral philosophy and his notion of the 'categorical imperative'.[28] This is the moral principle that individuals adopt and adhere to as a matter of free will and reason. It is a *categorical* imperative, because it applies to all rational beings unconditionally. That this is so is revealed by the proper use of one's rational faculty, as one cannot fail to adhere to the categorical imperative without inconsistency. The categorical imperative balances individualism with a commitment to what is good for all, in that individuals are all equally rational and so will choose how to conduct themselves from a series of options that are rational for all. When Kant turns explicitly to the question of war, he thinks we can use the standard of rationality imposed by the categorical imperative to deduce a set of shared rules or principles that must be adhered to in war and after the cessation of war, for a rational person would agree that war should be a rule-governed activity to ensure that innocent people are not injured, and so on.[29, 30] These rules or principles must be compatible with maxims formulated to ensure that, if other nations were to follow them as if they were parties to a contract, we shall eventually get to a point where there is perpetual peace among states and standing armies are abolished. These principles encourage restraint and respect for one's enemies and, according to Kant, rule out wars of extermination and subjugation, while also imposing strict standards on the means of warfare and post-war operations.[31]

Yet perhaps the most influential and systematic treatment of contract theory as it pertains to war can be found in John Rawls's *Law of Peoples*,[32] which draws extensively on Walzer's *Just and Unjust Wars*.[33] Rawls's argument is linked to his general approach to liberal values that is presented and defended in his other works, including *A Theory of Justice*[34] and *Political Liberalism*.[35] Unlike Suárez and Pufendorf, Rawls writes not of an implicit war contract, but of a hypothetical social contract. Rawls takes a two-level approach to international relations and war: he starts with domestic social contracts and then overlays that same sort of contract structure onto the international realm. For the sake of this chapter, we begin with the second (international) original position, which parallels the domestic original position.[36] In this second original position, representatives of each nation agree to a basic international structure and to principles governing it, all from behind the veil of ignorance. To be behind the veil of ignorance is to be deprived of information concerning the people whose fundamental interests one represents, including details of territory, population, its relative social, political and economic strength, and so on.[37] However, let us grant that the representatives know that conditions are favourable for constitutional democracies and that they represent liberal societies.[38] The second original position and the second-level veil of ignorance are heuristic devices that have been designed to ensure the impartiality of the parties involved. Under the international contract, Rawls argues that it would be rational for states to agree to just war principles based on the attempts of the contracting parties to maximise self-interest under the conditions of ignorance outlined above.[39] Rawls's hypothetical social contract not only holds a great deal of explanatory power, but will also appeal to many because of its pluralistic nature. The Rawlsian method allows people of different metaphysical and ethical worldviews to come together and embrace just war theory for very different reasons.[40, 41]

This is not to say that Rawls's hypothetical social contract, which leads him to just war principles, is without its problems or critics. Yvonne Chiu notes that, whereas traditional contract theories such as those from Hobbes, Locke, Rousseau and Rawls – which propose the existence of some form of contract between members of a single society engaged in the cooperative enterprise of promoting survival, greater wellbeing and propriety – may make sense at the initial domestic level, they make less sense when mapped onto the global arena.[42] In other words, Chiu rejects the carrying over of the domestic social contract model to the international realm and takes issue with Rawls's intrastate model. Rawls, she says, does not adequately explain why the contract is between 'peoples' rather than individuals.[43] In her view, representative individuals also need to be parties to the contract, even if it is primarily between well-ordered societies that are representative of their people. Chiu entirely dismisses the idea of a world government because it is both non-existent and highly improbable in the foreseeable future. As a remedy, she puts forward her own model, 'combat contractualism', which draws on the principles that underlie Rawls's model, but parts with it at the point of mapping the social contract onto the international arena. Chiu's combat contractualism denotes a pact, just short of a contract, between individuals whose agreement is mediated by the state as opposed to a contract directly between individuals and larger political units (as per Rawls).[44] That is, under her combat contractualism, the individual is not as subordinate to the state as under Rawls's account but retains a greater degree of autonomy. Nevertheless, Chiu argues that parties to this form of contract, behind a veil of ignorance not all that dissimilar to that of Rawls's (in which people are aware of the existence of other societies), would form an indirect agreement on fair terms of cooperation that can be stated in just war-like terms.[45] The general argument, then, is that regardless of the social contract tradition from which we draw and the specific division of responsibilities between military and state, we arrive at a form of just war theory. Let us now consider the core elements of such theory.

The Traditional Just War Duo: (i) *Jus ad Bellum*

Regardless of the way in which just war theory is grounded or underpinned, it has developed into a number of (relatively) distinct principles. According to proponents of traditional just war theory, the 'moral reality of war' is divided into two equally important dimensions. As Walzer says, war is therefore judged twice: 'first with reference to the reasons states have for resorting to war, secondly with reference to the means they adopt'.[46] He notes that the first sort of judgement is adjectival in character: we typically say that a particular war is just or unjust. The second is adverbial: we say that war is fought justly or unjustly.[47] Latin-writing medieval philosophers, theologians and jurists, on the other hand, have made the matter one of prepositions, distinguishing between *jus ad bellum* (justice of war) and *jus in bello* (justice in war). Together, they are the traditional 'just war duo'. However, in recent times there has been considerable discussion about dividing the moral reality of war into three parts, with the addition of yet another preposition: *jus post bellum* or justice after war. This dimension is detailed in Chapter 8. For the moment, we will discuss the six *jus ad bellum* principles, followed by the two *jus in bello* principles, before moving on to consider the relevance that they have for the contemporary problems associated with drone employment.

The primary *jus ad bellum* principle is that of *just cause*. A state that wants to be considered justified in their use of military force must have a rightful cause. Some argue that with more modern forms of warfare, we need to have an especially compelling reason for resorting to violence as modern weapons can have particularly terrible consequences for humanity. Of course, as Bruno Coppieters et al. note, we need some way of differentiating the good and just reasons for going to war from the not so good and clearly unjust reasons for going to war.[48] There is no easy way to go about this. There have been many heated debates concerning what warrants warfare. Historical writings on just war put forward a number of separate and somewhat different 'just causes' that are no longer recognised as such today. Aristotle, for example, argued that states have a right to go to war to gain an empire, provided that it would benefit everybody (including the conquered) and that it would not become so large and wealthy as to encourage further wars. Even more controversially by today's standards, Aristotle also thought it just to engage in warfare to acquire slaves for one's community, provided that such slaves were already 'naturally servile'.[49] Cicero, for his part, rejected this cause but argued, just as concerningly, that troops were justified in fighting for (Rome's) honour and glory.[50] These 'just causes' obviously do not feature in any modern form of just war theory.

There is, however, a venerable notion that defence always constitutes a just cause. This is generally accepted, at least in theory, as the only remaining legitimate cause for resorting to war in the contemporary just war framework.[51, 52] The term 'defence' is here to be broadly construed. In Walzerian language, defence would be construed as any action undertaken in response to some form of unwarranted (and often ungoverned) aggression. However, Stephen Coleman notes three specific forms of defence that may constitute a just cause:

least controversially, the defence of one's own country from aggressive attack; slightly more controversially, the defence of another country from aggressive attack; and rather more controversially still, the defence of innocent citizens from the aggressive attack of their own government.[53] The first case is a case of self-defence and hence it is fairly straightforward in standard cases. While perhaps not ideal,[54] in our current state of affairs, we have a system of independent states, each with discrete territorial integrity, political sovereignty and a people that it is responsible for protecting.[55] At the very least, each state in this system has a duty to protect its own means of government and the interests of its citizens. Thus, if one state is threatened by another's use of force, it has both a right and an obligation – as in the military-state contract developed earlier – to respond in kind, provided that, in doing so, it does not violate other just war principles.[56]

With self-defence typically granted as a just cause for engaging in warfare, it is often thought that the case can be readily extended to the defence of other states. For instance, the argument is often made that in the absence of a successful international police force, the only way to prevent international society from falling into a state of universal tyranny is to distribute 'policing powers' among the community of states.[57] It is sometimes argued in the military ethics literature that, if a state can be satisfied that another is facing unwarranted aggression, and needs assistance, then this state has a right and maybe even a responsibility (constrained by the primary responsibility it has toward its own citizens) to aid militarily the state under threat, and that doing so represents a just cause. The reasoning here is that if there is a breakdown of the international community of states, then the welfare of the individuals that constitute these states would be threatened. A more complicated question, which must await more comprehensive treatment elsewhere, concerns what obligations states owe to those under attack from their own government (these repressed individuals tend to be secessionists). Self-interested members of the international community of states, who stand to benefit from a stable international arena, would seem to owe some sort of obligation to the vulnerable population of an aggressive state and little, if anything, to the institutions of the aggressive state itself. From a social contract perspective, it can be said that once a state fails to meet the requirements of the contract it holds with its citizens by harming them in a systematic and gross way, it can no longer be considered a legitimate sovereign state.[58] Hence, it may be argued that a state would not be infringing on another's sovereignty by intervening, and that defending oppressed citizens is, in principle, another just cause that war may serve.

Apart from trying to determine what constitutes a just cause, another extremely important question that arises in relation to the notion of defence and its relation to drone attacks is whether the state is ethically permitted to employ weaponry in 'anticipatory' attacks, including both 'preemptive' and 'preventative' attacks. There is considerable confusion in the literature concerning the difference between these two terms, so we must be careful to differentiate the two – especially since one sort of attack is much more easily justified than the other. Drawing on Rodin and McMahan's distinction, preemptive attacks, in the context of this discussion, will include cases where enemy aggression is *judged to be imminent*, whereas preventative attacks will include cases where enemy aggression is *expected in the not-too-distant future*.[59] The chief question here, with particular relevance to targeted killing, is whether it is ethically permissible for a state to strike first in order to deter aggression from another state. According to just war theory, the answer is a conditional 'yes'. Walzer argues that states may use military force in the face of threats of war, whenever failure to do so would pose a serious risk to their territorial integrity, political independence, or the lives of those the state serves to protect.[60] This is arguably too permissive in that the costs of acting preemptively will often outweigh those of inaction. Yet, it seems reasonable to suggest that a state does not have to provide its a enemy with an opportunity to cause harm to its citizenry before taking action. Just as most would grant that it is justifiable to kill a gun-wielding maniac who poses a life-threatening risk, it can be argued that a state has a moral right to engage in a preemptive attack to protect its citizens. This right, however, is conditional on the degree of epistemic certainty held about the imminence of the attack. This is why preventative attacks are so much more difficult to justify than preemptive attacks: it is often impossible to know, until clear and present danger presents itself, how a state intends to act.

The second *jus ad bellum* requirement for just warfare is *right intention*.[61] The right intent principle is often read as saying that a state and its warfighters should not 'enter a war with hate in their hearts'.[62] This is an oversimplification, but as this reading might suggest, the principle does direct much attention to the subjective dimension of the use of armed force.[63] Coady argues that its primary purpose is to ensure that war's aims are grounded strongly in the just cause condition.[64] This is important because a state may very well have a just cause where that is but one of many competing interests. In some cases, it may even be that the intention of going to war has little or no relevance to the presence of a just cause. As evidence of the last two points, consider the multitude of intentions with which Coppieters and Kashnikov suggest wars are often waged: defending the state and others (as above); seeking glory or revenge; encouraging colonial expansion; promoting a particular religion; and, among many others, preventing the strengthening of an adversary's military.[65] From a moral point of view, some of these intentions are clearly more worrying than others. The extent to which they should be factored into the principle of right intentions depends on the strictness of the interpretation of the principle itself. The correct interpretation should encourage a close examination of the state's intentions to find the primary just cause intention, but it should also allow less 'pure' (although not abhorrent) intentions to be factored into the resort to war if they further legitimate military action in some way, so long as the primary just cause intention is allowed to dominate.[66] Of course, a state's intentions are not easily subjected to external examination. There are inevitable epistemological problems: no one can in all certainty know the intention of another.[67] Yet intention can in some sense be inferred from actions or results. For instance, the presence of wanton killing, destruction and humiliating domination give some indication of the likely absence of right intention and the presence of an intention to merely seek revenge,[68] which is not a morally acceptable motive for waging warfare.

Third, a morally justified decision to wage warfare must not only be supported by a just cause and right intent, but must also be undertaken or led by an individual or body that has the *right (or proper/legitimate) authority* to do so (which will be referred to as 'right authority'). The principle of right authority exists to regulate and legitimate any use of force and ensure that the power to authorise the use of force and the deployment of weapons rests only with certain institutions or persons. As argued in the previous chapter, the primary obligation of a state organised under the social contract is to provide effective and efficient protection from foreign enemies and to avert the risk of anarchic violence rising up from within the society itself. Members of such societies cede to the collective national institutions their right to use force – a right that they would otherwise possess in the state of nature. They do so, as Dan Zupan notes, in order to reap the rewards of a more civilised society, namely security from violence and harm.[69] This means that, in all but a limited number of cases, contracting parties essentially agree to refrain from waging private wars and accept that the right or responsibility to resort to violence is one reserved exclusively for the state as the apparatus of the collective.[70]

The task, then, is to identify whom or what institution, if any, qualify as holding right authority. Historically, princes, kings, popes and emperors have made claims to the authority to make war. For instance, a monarch would claim that they hold some sort of divine authority on earth and hold the power to wage war (though, of course, this would be disputed by the Pope).[71] However, over the last few hundred years, assumed authority has shifted from autocratic monarchs with their supposed divine authority, to the collective will of those who stand to be affected by the benefits and harms commonly associated with warfare.[72] Most theorists now think that in the United States, right authority rests with the people's democratically elected representatives in the Congress, who constitutionally hold the power to declare and/or authorise war.[73] Others think that war cannot be considered just for any country unless initiated in accordance with the guidelines or permission of some international body, like the United Nations. As with just cause, the principle of right authority is not entirely clear. It does, however, rest on a firm tradition that demands that decisions concerning war be made non-arbitrarily and that such considerations be based on the interests of societies.[74]

For war to be considered just, it must also meet a fourth requirement: the *likelihood of success* principle.[75] By placing a burden on those with decision-making authority to ascertain whether the planned war effort stands a reasonable chance of success, the principle serves to ensure that war is not unleashed capriciously or, as Coady remarks, 'without regard for outcomes'.[76, 77] While this presents as a fairly straightforward principle and serves to limit the waste of human life and economic resources, the ambiguity surrounding what the term 'success' actually entails means there are some complexities concerning its application. The concern is that success might be conceived in an operationally focused manner, which defines it as achieving set military tasks, whether that is bombing key buildings, stopping the enemy from crossing a certain line on a map, or so on.[78] However, operational success can be achieved without attaining the more important strategic goal, i.e. the just cause, to which operational successes are supposed to contribute. Hence, the problem is overcome when understood in just war terms, because success involves the attainment of the just cause that was the primary reason for initiating the armed conflict, and anything less (like mere operational success) would be a failure. Yet another concern is the ambiguity associated with the terms 'reasonable' and 'likelihood',[79] with the standard for anticipated success relating to how seriously each state and its military institutions take their obligation to be effective and efficient in providing defence under the military-state contract. Here, the more prudent states, or those with more openly engaged citizenries, will be more likely to demand higher anticipated levels of success than less prudent states or those with less openly engaged citizenries. However, perhaps the most serious concern is that, even if a state can discern a reasonable likelihood of success from an unreasonable one and recognise what true success actually involves, it might overestimate its abilities or underestimate its enemy's potential. Research undertaken by Dominic Johnson found that states and their citizens often exaggerate their ability to control the outcomes of conflict and entertain deceptive illusions concerning their chances of success, which is a problem magnified by technology.[80]

There is a fifth principle that is closely related to the likelihood of success principle: the *principle of proportionality*. Proportionality has two faces as a moral principle focused on restraining warfare. At the *jus ad bellum* level, the broad purpose of this principle is to ensure that the overall harm done as a result of warfare to life, property and values, will at least be balanced by the lives, properties and other valued things that will be preserved or protected.[81] Hence, it differs from the likelihood of success principle in that it is universal in nature.[82] Properly applied, not only does it aim to reduce the harms to all the belligerents involved, but it also accounts for the impact on civilians, the environment and so on. Paul Ramsey explains that proportionality is achieved when 'more good will be done than undone or a greater measure of evil prevented'.[83] William V. O'Brien holds a similar view and writes that 'successful recourse to armed coercion in pursuit of the just cause must outweigh the probable evil that the war will produce'.[84] These quotes quite reasonably suggest that, if the anticipated total good outweighs the total bad or evil to a probability of more than 50 per cent, it would be just to wage warfare, all other principles being met. That said, in applying the proportionality principle in practice, we come up with a few problems. Most are linked to the ever-present measurement problem that plagues war. The 'fog of war', along with what might be termed the modern 'haze of information overload', often precludes us from making exact probabilistic determinations. However, there are also many instances of war, both real and hypothetical, that are clearly proportionate or disproportionate. Nick Fotion argues that a strong state going to the defence of a weak state under overwhelming attack exemplifies proportionality,[85] while Coady argues that a nuclear bombardment in response to the persistent invasion of a state's fishing grounds exemplifies great disproportionality.[86] Of course, the practical value of a principle is determined most significantly by its capacity to distinguish and deal effectively with the hard and more marginal cases, not the easy ones. The proportionality principle also succeeds in this sense, reminding those applying just war theory that when the balance between total good and evil is particularly fine-grained, application and adherence to all other just war principles must be particularly strong if the justification for war is to be robust.

Finally, the fulfilment of *jus ad bellum* principles requires that warfare be a *last resort*. The presumption behind this principle is that since warfare is an evil in itself, it should be avoided insofar as there exist other potentially less damaging options. This idea can be traced back most notably to Cicero. He wrote that Roman law requires that a certain period of time elapse between the demand for reparations and the initiation of armed hostilities.[87] In fact, the Romans established a very strict protocol to be applied in the lead up to war, leaving 33 days for a peaceful resolution of the relevant problem.[88] Despite its ancient origins and changes in the tempo of war, the principle of last resort still holds a great deal of contemporary relevance. It serves to ensure that the emphasis is on resolving quarrels and disputes in ways that do not involve resorting to armed conflict, thus avoiding the powerful destabilising potential that always comes with violence.[89] In other terms, it obligates political leaders to make use of the nonviolent instruments that maintain relationships, yet which are still capable of achieving political objectives.[90] According to Coady, this can involve a resort to 'coercive diplomacy', which may include certain forms of deprivation or threats, a refusal to proceed with economic/political aid or an imposed limit on intellectual and cultural exchange, and so on.[91] Interestingly, as a form of intellectual exchange, the sharing of military technologies between states can defuse strong tensions (but it can sometimes create others). The key point, however, is that the war must be a genuine last resort and every other reasonable step must be taken first.[92] On the one hand, the principle of last resort encourages caution and prudence. It also reminds us of our moral duty not to unduly harm others and that the use of armed force may inspire an opponent to use retaliatory force. On the other hand, the principle reminds us that the prompt resort to armed force may, in some cases, be legitimate. For example, when facing clear danger and there are no other more advantageous options, it may not be prudent to delay the decision to go to war, since doing so may only increase the total level of harm incurred.[93]

Taken together, the six *jus ad bellum* principles place limits on when it is morally acceptable to wage warfare. They place a serious burden on any political decision maker who might be considering sending the state to war. If such a war is to be just, they must go through the substantive procedural steps and must meet the relevant moral requirements before launching war, regardless of how just the cause may seem. While interpreting and applying all of these principles in what is often a quickly unfolding hostile environment may seem a tricky, even impossible, proposition, all of these principles must be adhered to in order to meet the just war standard.[94] Indeed, the argument here is that it is critical for leaders to face this at-times formidable set of hurdles, not just for the good of the enemy and the international community, but also for the good of those whom they represent.

The Traditional Just War Duo: (ii) *Jus in Bello*

Once a state has decided that it is right and necessary to go to war, one of the most troubling problems it faces in deciding how to fight centres on the issue of whom they should use their weapons to target and whether these weapons are capable of the requisite degree of discrimination. Accordingly, states and their military institutions ought to look to the principle of *discrimination* for guidance on the matter. Cady states that, taken in a very broad sense, the principle holds that participants in war have an obligation in all but some very rare and limited circumstance to discriminate between appropriate and inappropriate targets of military destruction and harm.[95] This distinction, as will soon be shown, is based on the nature of the targets themselves, the activities they are involved in and their wider role in supporting the military. Historically, most parties to war have agreed in a contractual fashion that the destruction and harm caused to life and property beyond the class of combatants is an evil to be minimised when and where possible. As such, the lives and property of innocents, along with certain pieces of collective property, should not be deliberately targeted in any form of warfare. Walzer quite aptly notes that this principled restriction on the reach of warfare is the only thing that distinguishes war from common murder or massacre.[96]

Despite the fact that there is some agreement concerning both the practical and moral importance of protecting innocent persons during wartime, civilian casualties have been a notable feature of just about every significant conflict of the twentieth century and are likely to remain so for quite some time into the twenty-first century, even with the employment of unmanned systems and other emerging military technologies. Minimising the deaths of civilians, which are often referred to as a form of 'collateral damage' in the largely disconnected world of high-technology warfare, is the object of the subordinate but foundational rule of *noncombatant immunity*.[97] This rule offers protection to truly innocent civilians, but only by way of differentiating *combatants* from *noncombatants*. Thomas Nagel writes that we distinguish one from the other on the basis of their 'immediate threat or harmfulness', though he goes on to note that this line is far from sharp.[98] Yet the moral requirement is that there be some way for soldiers to recognise those who qualify as such. Standard practice is to consider those in uniform to be targetable combatants. However, under existing legal and moral conventions, wounded soldiers are not considered combatants and, as such, are not considered targetable. Conversely, Anthony Hartle points out that guerrilla fighters and insurgents *do not* wear uniforms but are clearly combatants.[99] The same can be said of private military contractors. Therefore, uniforms may provide an indication of combatant status, but cannot be considered the only signifier.

Some extend 'combatant' status to anyone in the 'chain of agency'; that is, anyone who is engaged in an activity inseparable from warfighting, including the manufacture of supplies.[100, 101] On a practical level, this makes it even more difficult for soldiers to differentiate between combatants and noncombatants. However, it is not always as difficult as it may look at first sight. There will obviously be many clear cases. Armed individuals that are aggressively pursuing you are clearly appropriate targets, while Cady notes that children (perhaps with the exception of child soldiers) and the infirm are clearly inappropriate targets of the means of war.[102] Yet, in the modern age, there are other problems for noncombatant immunity. Battlefields are no longer as clearly demarcated as those of Agincourt or Crécy. James Turner Johnson also alerts us to the fact that more people are involved.[103] This is partly because weapons technologies have become increasingly sophisticated, with warfare moving slowly into more populated areas (a problem in itself), and partly because developing states are strengthening their military forces. Overall, this has made it more difficult for soldiers to discern combatants from noncombatants, even once reliable criteria for differentiating the two have been established.

This difficulty seems to confirm Walzer's point that the lives of noncombatants are often only at risk because they are in close proximity to violent battles being fought by others.[104] The principle of discrimination requires that care be taken not to harm these noncombatants, but exactly how much care must we take? This is really a question of what level of risk soldiers should be required to assume. Again, these are questions that will be reflected upon in upcoming chapters, but there is one related question that must be addressed at this preliminary point. Is it just to destroy a military target if the destruction of that target might somehow collaterally injure or kill noncombatants? For instance, can we bomb an arms depot or an airfield if injury or death may be caused to innocents who unfortunately happen to be in the vicinity of these targets when the bombing takes place? Walzer explains that, in an effort to resolve these sorts of questions, Catholic casuists of the Middle Ages – who were committed to the just war tradition, yet equally committed to defending their states' interests and war efforts – put forward the doctrine of *double effect*.[105] This doctrine is potentially a way of reconciling the absolute prohibition against targeting enemy noncombatants with the legitimate conduct of military operations which serve to protect the interests of a state's own citizens, most of whom will also be noncombatants. The thrust of the traditional notion of double effect is essentially that it is permissible to perform an act that may have 'evil' consequences (principally, the death of noncombatants), so long as the following four requirements are met: 1) the act is good in itself; 2) the direct effect is morally acceptable; 3) the intention of the military actor is good; and 4) the good effect is substantially good enough to compensate for the evil effect.[106] The crux of the argument lies in the third clause: the matter of intention. Importantly, Walzer argues that good intention is not enough to justify such an act and that active measures must be taken to ensure that any potential evil is minimised.[107] Adding this extra premise to the third clause makes sense, as simply not intending the deaths of noncombatants seems a condition too easy to meet, especially if we grant that few soldiers enter war with the explicit aim of harming innocent people.

Very closely related to the notion of double effect – so close, in fact, that it is one of the conditions governing its invocation – is the *proportionality* principle. In the previous section it was noted that this principle, when used as a standard of *jus ad bellum*, requires that the anticipated moral cost of fighting a war not be out of line with the expected moral benefits. Here, when applied to the *jus in bello* domain, considerations of proportionality are to be used to determine the permissibility of specific actions in the prosecution of a war. In the words of van Damme and Fotion, the philosophical conception of the principle does not change, only the context in which it is applied.[108] In the *in bello* context, the proportionality principle is generally taken to be well within the realm of consequentialist theory, inasmuch as it is thought that only consequences – more precisely only welfare consequences – count in determining whether a particular action is proportionate to the good sought.[109] Unsurprisingly, then, most versions of the principle say that, in making operational decisions, military commanders and frontline soldiers ought to maximise good and minimise evil. Rawls, however, makes what is essentially the same point in a slightly more nuanced way. He says that 'the means employed must not destroy the possibility of peace or encourage a contempt for human life that puts the safety of ourselves and of mankind in jeopardy'.[110] With either formulation, there is a degree of speculation involved in determining how wide the scope of the proportionality principle is supposed to be. With a narrow scope, one would look at the likely goods and harms of a particular military operation and how it stands to affect those in the immediate vicinity. With a broader or more strategic scope, one would look at the harm done in a particular military operation and weigh that against the overall good to be accomplished by the war.[111] Either way, the principle reminds soldiers to consider the suffering of others and calls on them to restrain the harmfulness of war.

Just War Theory: Contemporary Relevance and Challenges

Having placed just war theory in context and reviewed its eight principles in preparation for making some observations about the specific problems posed by unmanned systems, we can now pause and reflect on the general moral guidance that the theory provides. It would seem that much of just war theory's moral strength rests in the following principles (which tend to be widely accepted): defence of the innocent, fairness, justice and moral restraint. In line with these principles, there are specific conditions that prevent an overeager interest in going to war and others that restrain war once it breaks out. Still, despite the historical weight of the tradition and the common sense guidelines that it provides, there exists a question as to whether just war theory is ultimately defensible as an appropriate way of thinking about the sort of contemporary warfare with which we are concerned.

Critics have suggested that war is not what it used to be and that it may be too much to expect just war theory to offer much in terms of useful moral guidance on the use of unmanned systems. Laura Sjoberg, for example, argues that a world with modern weapons, such as unmanned systems, would have been 'inconceivable for classical just war theorists'.[112] As a result, it would be difficult for the just war tradition to provide any insight into this world, especially because of the challenges posed 'by the attempt to discriminate between combatants and noncombatants'.[113] Sjoberg's solution is to radically reinterpret just war theory along feminist lines. Her morality of war would be based on 'relational autonomy, political marginality, empathy and care'.[114] In an even more critical tone, Laurie Calhoun writes that some academics have become trapped in a paradigm and that just war theory has little to do with the 'practices that go by the name of war today', going on to question whether those who spend their time 'rehearsing the standard list of just war theory requirements' are capable of contributing in any meaningful or constructive way to the resolution of the ever more vexing problems of modern warfare, which include those of unmanned warfare.[115] While offering no real alternative, Calhoun goes on to argue that, rather than spending entire careers defending the just war paradigm, academics should acknowledge the anachronism of just war theory, cease rehashing texts written centuries ago and begin, at last, to consider the 'actual practices of war in the twenty-first century'.[116] These are but two proponents of the commonly heard argument that, as Elizabeth Anscombe summarises in her refutation of pacifism, 'the old conditions for a just war are irrelevant to the conditions of modern warfare, so that must be condemned out of hand'.[117]

As proponents of this general argument, Sjoberg and Calhoun pose a genuine methodological question. Should we view unmanned warfare as merely one more area of application for just war theory or is it, as Luciano Floridi puts it, so 'disruptively novel' that it calls for new theoretical developments and perhaps a new macro ethic?[118] Influenced by his work in the field of information ethics, Floridi tends to think that instead of updating our old theories with 'more and more service packs',[119] it might be time to redevelop them altogether, just as a programmer would rewrite software. However, it will be argued in the remainder of this book that, while some questions may initially seem to invite new theorising rather than the adaptation of more traditional ideas, it is not the case that this applies to just war theory and the use of unmanned systems. Those who reject just war theory's relevance are yet to point to a concern about unmanned systems that cannot be addressed from within the tradition. These systems may have changed the *character* of warfare in morally relevant ways, but it will be shown that the *nature* of warfare has not changed so significantly that it justifies developing a new moral framework. The matter here is not one of overcoming the 'defects' that are supposedly present in the just war framework itself, but is rather one of finding an appropriate way of addressing and engaging with the concerns associated with the use of unmanned systems from within an already existing and sound framework. It will be argued that the challenge consists in carefully and critically interpreting as well as applying existing just war principles to the complexities of unmanned warfare. This has historically been the case when other emerging military technologies with disruptive potential have shown operational promise and have become the focus of increased attention among scholars of the ethics of war and weapons proliferation.

Consider the debate that began with the invention of chemical and biological weapons and has since expanded to cover nuclear weaponry. It was often said that the existence of these weapons made it overly optimistic to think that traditional moral principles circumscribing warfare, like those found in just war theory, could adapt or continue to offer any sort of practical guidance on the just resort to armed force and the just conduct of any subsequent battle. It was thought that these weapons could not be used in anything but an unrestricted fashion, breaking all the rules of warfare, rendering just war theory obsolete. However, this was far from the truth. Just war theory has continued to be regarded by many theorists as relevant, applicable and able (at least in essence) to accommodate recent changes in warfare and weaponry because its principles require us, as Walzer states,[120] to 'condemn even the threat to use [imprecise] nuclear weapons'. Richard Norman's *Ethics, Killing, and War*,[121] Brian Orend's *War and International Justice*,[122] William V. O'Brien, John Langan's *The Nuclear Dilemma and the Just War Tradition*,[123] Robert L. Phillips's *War and Justice*[124] and a host of articles commonly found in *Ethics* or *The Journal of Philosophy and Public Affairs* also defend just war theory and argue that its principles can be consistently applied to nuclear warfare and, indeed, that it represents the only sound way of thinking about the moral problems of such warfare. Even if a large Hiroshima-style attack can never be justified, just war theory can show us precisely what is wrong with it. The absence of nuclear warfare, despite many states continuing to hold large nuclear arsenals, might also suggest that the considerations enshrined in just war theory have been successful in averting a 'nuclear winter'. More recently, with the rise of global terrorism and an increased international interest in information warfare or, more specifically, cyberwarfare, academics have again found merit in the application of just war theory. In each case, just war principles have been adapted to serve as useful guideposts to the resort and conduct of these novel forms of conflict and warfare.[125]

Yet the fact that just war theory can be extended to cover these forms of warfare does not, in and of itself, confirm that just war theory is the right moral theory on which to base our assessment of the ethical issues associated with the use of unmanned systems. In other words, the question of adequacy still remains. To demonstrate that just war theory is a suitable moral theory for the task at hand, we must consider what the requirements are of a good system of moral guidance. James Turner Johnson suggests that when we ask about the adequacy of a particular system of moral guidance, we are in fact asking three separate but in some sense related questions:

> Is the system "right"? That is, does it correspond to the moral values of the culture or community in which it has come into existence and in which the identity of the moral agent has been shaped?
>
> Does the system provide [a sufficiently robust] conceptual framework for moral analysis and judgement?

> Does this system represent a bridging of the gap between the ideal and possible? That is, does it produce practical moral guidance as well as identify the relevant moral values for the situation at hand?[126]

To sustain the claims made about the contemporary relevance of just war theory and its central place in morally evaluating unmanned warfare would be to answer all three questions in the affirmative. And we can indeed answer each of these questions positively. This will be addressed only briefly here, since the principal aim is not to offer a general defence of just war theory and its applicability to modern technological warfare – others have done this quite convincingly elsewhere[127] – but rather to offer a more limited defence of the theory as a useful tool for accessing and analysing the complicated problems associated with the employment of unmanned weaponry.

Let us start with the first question: does just war theory correspond to the moral values of the community it serves? We can safely say that just war theory affirms a common set of moral perceptions or, more particularly, the common moral language of the Western world, which (let us not forget) employs the greatest number of unmanned systems. As Johnson has recently noted, people in our culture cannot seem to escape thinking about the morality of war in terms of the just war paradigm.[128] This, in itself, provides a reason for employing a just war framework as a descriptive ethic. Of course, as mentioned in the second chapter, many non-Western states also utilise or are in the process of developing unmanned platforms, and even the states that do not have such plans stand to be affected by the actions of those that do. For this reason, it is important to note that just war theory has its roots in intellectual traditions from around the world and, while primarily a Western ethic, it can also be universalised under contract based accounts and has been applied across cultures.[129] This means that in the absence of an alternative, just war theory can be adopted as a guide to actions by all those utilising unmanned systems, meaning that there would be an inherent reciprocity involved. Any action that can legitimately be undertaken by a Western state can also be legitimately undertaken by a non-Western state under the same circumstances, not only improving fairness, but also giving just war theory the best chance of limiting the damage done by war. This should dispel the claim that just war theory is a heuristic designed to suit the needs of the West and justify its use of unmanned weaponry.

Concerning the second question of whether the just war framework is conceptually adequate for making the necessary moral analyses and judgements of unmanned systems, we must seriously ask what else might replace the just war conception and the principles of just cause, right authority, right intent, likelihood of success, proportionality, last resort and discrimination. As already argued in the first section of this chapter, because of the enduring value in just war theory, it is not simply a matter of replacing these principles with straight consequentialist concepts, nor of providing alternative interpretations of others such as the 'common good' or codes of justice such as *lex talionis*. Mark Evans rightly states that the challenge to critics is whether, insofar as we wish to think normatively about the use of unmanned systems, we can conjure up principles and put them into any form of coherent structure without leading us back to some form of just war theory.[130] There may be other concepts that could be formulated to yield more useful or determinate results than the just war concept, but as Johnson remarks, this is equivalent to saying that 'there may be life in the fifth dimension; we simply cannot know because our knowledge is limited by the circumstances of our existence'.[131] As such, it would seem that the conceptual moral framework provided by just war theory is the *best* we currently have to deal with the contemporary challenges of warfare and the emergence of new military technologies such as unmanned systems.[132] This is not to give the impression that just war theory is capable of handling *all* the challenges posed by unmanned systems. However, the burden falls on Calhoun, Sjoberg and other critics of just war theory to demonstrate and explain its conceptual inadequacies in a concrete and substantial way.

In regards to the final question concerning what degree of moral guidance is provided by just war theory, it is critical to point out that both the practical and the theoretical have come together to make just war theory what it is today. On the one hand, just war theory reflects a commitment to the need for practical relations between states of the international community. On the other hand, it reflects a more theoretical commitment to a better state of world affairs, in which the resort to armed conflict is a less common occurrence. As a whole, just war theory presents us with a system of moral guidance that allows us to determine whether a particular decision to employ unmanned weaponry is justified or not and whether it can or has been used appropriately in battle. However, it is important to bear Aristotle's warning in mind: a moral theory can only be as clear as its subject matter allows.[133] Precise moral judgements are often blurred by war's messiness and, as such, some just war principles have unclear – or even elastic – boundaries that can be interpreted in a number of ways.[134] While there are some cases where the justice of a particular conflict or action is so opaque that neither side should even consider entering into an armed conflict, Johnson has made the argument that the elasticity of just war theory is important in the sense that there is often more to be learned when moral principles are open to diverse reasonable interpretations than when they are not.[135] The argument here is that it is a special merit of just war theory that its principles are open to fair interpretation, making the theory more useful as a diagnostic for modern conflicts. This is what Nicholas Evans describes as a 'mechanism of self-disclosure and self-classification'[136] or, in simple terms, a tool which encourages moral reflection.

An application of the just war approach causes us to focus on what would appear to be some of the most important and fruitful questions relating to the employment of unmanned systems. Will unmanned systems provoke preemptive attacks? Might states go to war with the unjust intention of testing new and operationally unproven platforms? Will these systems blur the distinction between combatant and noncombatant to too great an extent? How will their use impact upon the proportionality calculus? Will they make the decision to resort to war an easier one? Can they meet the standard of discrimination? Such questions will constitute the focus of the remaining chapters of this book. In answering them, we must remember that because just war theory is open to a degree of interpretation and because unmanned warfighting technologies are constantly evolving, we cannot always assume that the use of a technology that is said to satisfy just war principles will always be moral; nor can we assume that the use of a technology that does not satisfy these principles will be forever immoral.[137] Nevertheless, the contemporary principles of just war, while imperfect, are well attuned to the practical reality of armed conflict and the harsh realities of the battlefield in a way that other theories with more abstract moral principles are not. No other framework will yield the sort of clearly delineated questions raised above, and which we will need if we are to thoroughly analyse the ethics and efficacy of the use of unmanned systems.

It is also important to recognise that, because just war theory aims to bridge the gap between what is ideal and what is actually possible, the two central pillars of the theory are, as a matter of structure, considered logically independent. This means that for the purpose of the ensuing discussion, it is perfectly possible for an unmanned war to meet one of these set of standards without meeting the other. In other words, wars that fail to meet the standards of *jus ad bellum* may still be conducted in a justifiable manner, that is, in accordance with *jus in bello*. Wars that do meet the standard of *jus ad bellum* may also be conducted in an unjustifiable manner if they are not in accordance with *jus in bello*. While *jus post bellum* has not been addressed above, it too is logically independent in a similar way. So a war might be conducted justly, in accordance with *jus ad bellum*, but then fail to meet *jus post bellum* principles after the conflict. However, recent work on just war theory – most notably that of McMahan[138] – has called into question whether it is possible to meet any of these standards without first satisfying the just cause requirement of *jus ad bellum*.[139] In other words, McMahan challenges the traditional independence of the standards of just war theory. McMahan's contemporary challenge to the reigning theory is discussed in more detail in Chapter 7 in relation to the use of unmanned systems in radically asymmetric conflicts. The real value of just war theory, however, comes from the fact that it provides us with moral guidance when it is impossible to determine matters of objective justness, which are central to McMahan's thesis. For the moment, however, let it suffice to say that the traditional distinction between the different branches of just war theory remains highly relevant to the practice of war and the use of unmanned weaponry in our non-ideal world, despite any tensions that may exist between them.

Conclusion

Chapter 3 concluded that there was a carefully delimited duty to employ unmanned systems and, in this chapter, we have explored grounds upon which a call for restraint can be based. The first section contrasted just war theory with pacifism and realism as its foundational rivals and argued that it is more useful than these alternative theories, at least as they apply to the unmanned systems debate. It was said that just war theory could be philosophically underpinned by a number of ethical theories, most interestingly, those contract theories discussed by Kant, Suárez, Pufendorf, Rawls and Chiu, and that regardless of the way in which just war theory is grounded, we are led to embrace just war-like principles. In the second section, we considered the first set of these principles: those of *jus ad bellum*. Under this standard, war must have a just cause, be waged with right intention, undertaken by a right authority, have a reasonable likelihood of success, be proportional and a last resort. We then considered the two principles that fall under the *jus in bello* standard, requiring warfare to be conducted in a discriminatory and proportional manner. The fourth and final section argued that, despite claims to the contrary, the classical just war framework remains the most suitable and robust tool we have for analysing the moral problems associated with the use unmanned systems. However, we must be especially careful to consider how the changing nature of warfare and warfighting technology may require us to rethink how we apply some aspects of the framework in the just war-centric aspects of the upcoming discussion of the operational, psychological and more philosophical objections to the use of unmanned systems.

Notes

1 While just war theory is often discussed as though it is a single theory, it is actually many theories and lines of traditional thought that are referred to as a coherent whole under the term 'just war theory'. At times, the term 'just war tradition' will be used here to refer to this body of thought and literature.

2 James T. Johnson, *Can Modern War Be Just* (New Haven: Yale University Press, 1984), 18.

3 Ibid.

4 Duane Cady, L., *From Warism to Pacifism: A Moral Continuum* (Philadelphia: Temple University Press, 2010), 17.

5 Ibid., 40.

6 Andrew Fiala, *The Just War Myth* (Lanham: Lexington Books, 2008), 163.

7 Douglas P. Lackey, *Moral Principles and Nuclear Weapons* (New Jersey: Rowman & Allanheld Publishers, 1984), 11.

8 Nick Fotion, Bruno Coppieters and Ruben Apressyan, 'Introduction', in *Moral Constraints on War: Principles and Cases*, ed. Bruno Coppieters and Nick Fotion (Lanham: Lexington Books, 2008), 10.

9 Hugo Grotius, *The Rights of War and Peace*, trans. A. C. Campbell, vol. 2, 24 (Washington, DC: M. Walter Dunn, 1901), 18.

10 Just war principles may be used as a guide in determining when exercising restraint is in the interests of the state, but these principles are sought out and perhaps adhered to for no reason other than the fact it is best for the state.

11 Cady, *From Warism to Pacifism: A Moral Continuum*, 32.

12 Laura Sjoberg, 'Why Just War Needs Feminism Now More Than Ever', *International Politics* 45, no. 1 (2008): 2.

13 Fotion, Coppieters, and Apressyan, 'Introduction', 20.

14 For some utilitarian accounts, see Richard Brandt, 'Utilitarianism and the Rules of War', in *War, Morality, and the Military Profession*, ed. Malhan Wakin (Westview Press: Boulder, 1981); R. M. Hare, 'Rules of War and Moral Reasoning', in *Essays on Political Morality*, ed. R. M. Hare (Oxford: Clarendon Press, 1989); Nick Fotion, 'A Utilitarian Defense of Just War Theory', *Synthesis Philosophica* 12, no. 1 (1997).

15 C. A. J. Coady, *Morality and Political Violence* (Cambridge: Cambridge University Press, 2008), 16.

16 Michael Walzer, *Just and Unjust Wars: A Moral Argument with Historical Illustrations* (New York: Basic Books, 2006), 131.

17 Nick Allen, 'Just War in the Mahabharata', in *The Ethics of War: Shared Problems in Different Traditions*, ed. Richard Sorabji and David Rodin (Aldershot: Ashgate, 2006); Norman Solomon, 'The Ethics of War: Judaism', in *The Ethics of War: Shared Problems in Different Traditions*, ed. Richard Sorabji and David Rodin (Aldershot: Ashgate, 2006).

18 United States Catholic Bishops, 'The Just War and Non-Violent Positions', in *War, Morality, and the Military Profession*, ed. M. Wakin (Boulder, CO: Westview Press, 1989).

19 James F. Childress, 'Just-War Theories: The Bases, Interrelations, Priorities, and Functions of Their Criteria', *Theological Studies* 39, no. 1 (1978).

20 Walzer, *Just and Unjust Wars: A Moral Argument with Historical Illustrations.*

21 At times, Walzer (as I read him) seems to argue from a communitarian social contract perspective, meaning that his discussion might also fall into the discussion below. See ibid., 53–54.

22 Stephen C. Neff, *War and the Law of Nations: A General History* (Cambridge: Cambridge University Press, 2005), 137.

23 Francisco Suárez, 'On War', in *Francisco Suárez, Selections from Three Works*, ed. Gladys L. Williams, Ammi Brown, and John Waldron (Oxford: Clarendon Press, 1944).

24 Ibid., 851–2.

25 Samuel Pufendorf, *On the Law of Nature and Nations*, trans. C. H. Oldfather (Oxford: Clarendon Press, 1934), 1294.

26 Ibid., 1325.

27 Immanuel Kant, 'Metaphysics of Morals', in *Kant: Political Writings*, ed. H. S. Reiss (Cambridge: Cambridge University Press, 1991), 167.

28 Peter S. Temes, *The Just War* (Chicago: Ivan R. Dee, 2003), 49.

29 Kant, 'Metaphysics of Morals', 168.

30 Some political theorists have argued that it is an error to place Kant in the just war tradition, but this remains a minority view and one that, even if valid, has little significance for the overall argument put forward here. For an example, see Howard Williams, *Kant and the End of War: A Critique of Just War Theory* (New York: Palgrave Macmillan, 2012).

31 Kant, 'Metaphysics of Morals'.

32 John Rawls, *The Law of Peoples* (Cambridge: Harvard University Press, 2001).

33 Walzer, *Just and Unjust Wars: A Moral Argument with Historical Illustrations.*

34 John Rawls, *A Theory of Justice* (New York: Oxford University Press, 1999).

35 John Rawls, *Political Liberalism* (New York: Columbia University Press, 2005).

36 Rawls, *The Law of Peoples*, 30–34.

37 Ibid., 32–33.

38 Ibid., 33.

39 Ibid., see 59–88 and 89–120.

40 Andrew Fiala, *Practical Pacifism* (New York: Algora Publishing, 2004), 79.

41 It is also compatible with the military-state contract mentioned earlier, as it could be argued that those under the first-level veil of ignorance would agree to establish such a contract.

42 Yvonne Chiu, 'Combat Contractualism: A Secular Theory of Just War' (University of California, 2007), 40–41.

43 Ibid., 39.

44 Ibid., 51.

45 Ibid., see page 70 et seq.

46 Walzer, *Just and Unjust Wars: A Moral Argument with Historical Illustrations*, 21.

47 Ibid.

48 Bruno Coppieters, Carl Ceulemans and Anthony E. Hartle, 'Just Cause', in *Moral Constraints on War: Principles and Cases*, ed. Bruno Coppieters and Nick Fotion (Lanham: Lexington Books, 2008), 27.

49 Brian Orend, *The Morality of War* (Peterborough: Boradview Press, 2006), 10–11.

50 Ibid., 11.

51 Johnson, *Can Modern War Be Just*, 19.

52 It should be noted that Rodin argues that the right to self-defence must ultimately rest upon individual rights and that when a large portion of the community is at stake, it may legitimately call on self-defence as a just cause for war. This follows well from the contractual account put forward earlier and is critical of standard views, which perhaps give too much value to sovereignty and too frequently permit wars over territory. The aim here, though, is not to defend either model but rather outline just war principles in their basic form. For more, see Rodin, *War and Self-Defense*, 139–40.

53 Stephen Coleman, *Military Ethics: An Introduction with Case Studies* (Oxford: Oxford University Press, 2012), 72.

54 There is still considerable debate concerning whether a world state is, in fact, ideal.

55 Cady, *From Warism to Pacifism: A Moral Continuum*, 34.

56 There are many old debates concerning whether both sides can have a just cause in war. This is not a matter I want to dwell on here, however, the common answer, advanced by Suárez and Vitoria, amongst others, is that it is impossible for both sides to have *objectively just* causes. This is the point that McMahan relies on in his challenge to traditional just war theory. I return to this challenge in discussing radical technological asymmetry in a later chapter. For the moment, note that it is possible for both sides to have *objectively unjust* or *subjectively just* causes. In both cases, just war theory serves the common interest in reducing the damage of war.

57 Coppieters, Ceulemans, and Hartle, 'Just Cause', 44; Walzer, *Just and Unjust Wars: A Moral Argument with Historical Illustrations*, 59.

58 Coppieters, Ceulemans, and Hartle, 'Just Cause'.

59 Rodin, *War and Self-Defense*, 113; Jeff McMahan, 'Preventative War and the Killing of the Innocent', in *The Ethics of War: Shared Problems in Different Traditions*, ed. Richard Sorabji and David Rodin (Aldershot: Ashgate, 2006), 170.

60 Walzer, *Just and Unjust Wars: A Moral Argument with Historical Illustrations*, 85.

61 The principle of right intentions can be subordinated to the principle of just cause, but it is discussed separately here for the sake of clarity, as is often done in such discussions.

62 Fotion, Coppieters, and Apressyan, 'Introduction', 12.

63 Coppieters and Kashnikov, 'Right Intentions', 73.

64 Coady, *Morality and Political Violence*, 99.

65 Coppieters and Kashnikov, 'Right Intentions', 73.

66 Jeff McMahan, Robert McKim and Thomas Hurka seem to agree with this point, or something similar to it, in making a distinction between 'sufficient' and 'contributing' just causes. See Jeff McMahan and Robert McKim, 'The Just War and the Gulf War', *Canadian Journal of Philosophy* 23, no. 4 (1993): 512; Thomas Hurka, 'Proportionality in the Morality of War', *Philosophy & Public Affairs* 33, no. 1 (2004): 41.

67 Cady, *From Warism to Pacifism: A Moral Continuum*, 35.

68 Ibid.

69 Daniel S. Zupan, 'A Presumption of the Moral Equality of Combatants: A Citizen-Soldier's Perspective', in *Just and Unjust Warriors: The Moral and Legal Status of Soldiers*, ed. David Rodin and Henry Shue (Oxford: Oxford University Press, 2008), 216.

70 Ibid.

71 Laura Sjoberg, *Gender, Justice, and the Wars in Iraq: A Feminist Reformulation of Just War Theroy* (Lanham: Lexington Books, 2006), 64.

72 Cady, *From Warism to Pacifism: A Moral Continuum*, 34.

73 The US President, as Commander-in-Chief of the armed forces, can order an emergency deployment of troops and start a military engagement, but cannot formally declare war.

74 Cady, *From Warism to Pacifism: A Moral Continuum*, 35.

75 Childress, 'Just-War Theories: The Bases, Interrelations, Priorities, and Functions of Their Criteria', 437.

76 Coady, *Morality and Political Violence*, 94.

77 For Augustine, God controlled war's outcomes 'according to his pleasure'. Thus, the outcome of war was not a matter for mere mortals. It was from this view he argued that the principle does not stand on its own. Augustine, *City of God* (London: Penguin Books, 1984), 216. For our sake here, let us side with Hugo Grotius in viewing it as a principle unto itself, albeit one that it is related to the principle of proportionality (just how is described below). Grotius, *The Rights of War and Peace*, 2, 24: 280–84.

78 Albert C. Pierce, 'War, Strategy, and Ethics', in *Ethics and the Future of Conflict: Lessons from the 1990s*, ed. Anthony Lang, Albert C. Pierce, and Joel H. Rosenthal (Upper Saddle River: Prentice Hall, 2004), 15.

79 Fotion and Coppieters, 'The Likelihood of Success', 101.

80 Dominic D. P. Johnson, *Overconfidence and War: The Havoc and Glory of Positive Illusions* (Cambridge: Harvard University Press, 2004).

81 Johnson, *Can Modern War Be Just*, 25.

82 It is worthwhile noting that the likelihood of success principle and the proportionality principle might give different answers about whether it is just to go to war. It is not overly difficult to imagine a scenario in which a state is capable of winning a war, but where fighting it would breach the military-state contract due to the costs being preventatively high. A similar point is made in: Coleman, *Military Ethics: An Introduction with Case Studies*, 82.

83 Paul Ramsey, *The Just War: Force and Political Responsibility* (New York: Charles Scribner's Sons, 1968), 195.

84 William V. O'Brien, *The Conduct of a Just and Limited War* (New York: Praeger, 1981), 27.

85 Nick Fotion, 'Proportionality', in *Moral Constraints on War: Principles and Cases*, ed. Bruno Coppieters and Nick Fotion (Lanham: Lexington Books, 2008), 125.

86 Coady, *Morality and Political Violence*, 96.

87 Paul Christopher, *The Ethics of War and Peace* (Upper Saddle River, NJ: Prentice Hall, 1994), 13.

88 Ibid., 13–14.

89 Johnson, *Can Modern War Be Just*, 24.

90 Pierce, 'War, Strategy, and Ethics', 15.

91 Coady, *Morality and Political Violence*, 91.

92 I use the terms 'genuinely and 'reasonable' as, in most cases, there will be some other option that the state could have utilised, but did not. The claim here is not that a state ought to go through *every* option, just those that the particularities of the case would reasonably allow.

93 Coppieters, Apressyan and Ceulemans, 'Last Resort', 140.

94 Pierce, 'War, Strategy, and Ethics', 15.

95 Cady, *From Warism to Pacifism: A Moral Continuum*, 37.

96 Walzer, *Just and Unjust Wars: A Moral Argument with Historical Illustrations*, 42.

97 Anthony E. Hartle, 'Discrimination', in *Moral Constraints on War: Principles and Cases*, ed. Bruno Coppieters and Nick Fotion (Lanham: Lexington Books, 2008), 171–2.

98 Thomas Nagel, 'War and Massacre', in *War, Morality, and the Military Profession*, ed. Malham M. Wakin (Westview Press: Boulder, 1986).

99 Hartle, 'Discrimination', 172.

100 Jeffrie Murphy, 'The Killing of the Innocent', in *War, Morality, and the Military Profession*, ed. Malham M. Wakin (Boulder, Colorado: Westview, 1986), 346. Coady, *Morality and Political Violence*, 111.

101 While questions of responsibility for war are expanded on in later chapters, this should strike readers as extending combatant status too far. Those who participate in activities that would not exist if it were not for warfare clearly hold some level of responsibility for conflict that their activity enables, but not all should be classified as combatants liable to attack.

102 Cady, *From Warism to Pacifism: A Moral Continuum*, 37.

103 Johnson, *Can Modern War Be Just*, 72–3.

104 Walzer, *Just and Unjust Wars: A Moral Argument with Historical Illustrations*, 152.

105 Ibid.

106 Ibid., 153.

107 Ibid., 155.

108 Guy Van Damme and Nick Fotion, 'Proportionality', in *Moral Constraints on War: Principles and Cases*, ed. Bruno Coppieters and Nick Fotion (Lanham Lexington, 2008), 159.

109 Ibid.

110 Rawls, *A Theory of Justice*, 379.

111 Van Damme and Fotion, 'Proportionality', 160–64.

112 This is in direct contrast with the evidence presented under the section on the ancient and mythological origins of unmanned systems in Chapter 2. Primitive forms of unmanned systems were conceivable to classical just war theorists and many would have been familiar with the myths involving automata, although this ultimately has little bearing on the argument here.

113 Sjoberg, 'Why Just War Needs Feminism Now More Than Ever', 2.

114 Ibid., 1.

115 Laurie Calhoun, 'Political Philosophers on War: Arguments inside the "Just War" Box', *The Independent Review* 15, no. 3 (2011): 447.

116 Ibid., 461.

117 Elizabeth Anscombe, 'War and Murder', in *War, Morality, and the Military Profession*, ed. Malham M. Wakin (Boulder, CO: Westview Press, 1986), 294.

118 Luciano Floridi, 'Children of the Fourth Revolution', *Journal of Philosophy and Technology* 24, no. 3 (2011): 230.

119 Ibid.

120 Walzer, *Just and Unjust Wars: A Moral Argument with Historical Illustrations*. In particular, see p. 276 where he argues against the apparent 'possibility that the new technology of war simply does not fit and cannot be made to fit within the old limits [of just war theory]'.

121 Richard Norman, *Killing, Ethics, and War* (Cambridge: Cambridge University Press, 1995).

122 Brian Orend, *War and International Justice: A Kantian Perspective* (Waterloo: Wilfrid Laurier University Press, 2000).

123 William V. O'Brien and John Langan, eds, *The Nuclear Dilemma and the Just War Tradition* (Lexington: Lexington Books, 1986).

124 Robert L. Phillips, *War and Justice* (Norman: University of Oklahoma Press, 1984).

125 For some interesting articles on terrorism, see W. Smit, ed. *Just War and Terrorism: The End of the Just War Concept?* (Leuven: Peeters Publishers, 2005); for a take on cyberwarfare, see Dipert, 'The Ethics of Cyberwarfare'.

126 Johnson, *Can Modern War Be Just*, 16.

127 William E. Murnion, 'A Postmodern View of Just War', in *Intervention, Terrorism, and Torture: Contemporary Challenges to Just War Theory*, ed. Steven P. Lee (Dordrecht: Springer, 2007).

128 James T. Johnson, *Ethics and the Use of Force: Just War in Historical Perspective* (Farnham: Ashgate, 2011), 76.

129 Coleman, *Military Ethics: An Introduction with Case Studies*, 66.

130 Mark Evans, 'In Defence of Just War Theory', in *Just War Theory: A Reappraisal*, ed. Mark Evans (Edinburgh: Edinburgh University Press, 2005), 220.

131 Johnson, *Can Modern War Be Just*, 17.

132 John P. Sullins, 'Aspects of Telerobotic Systems', in *Ethical and Legal Aspects of Unmanned Systems: Interviews*, ed. Gerhard Dabringer (Vienna: Institute for Religion and Peace, 2010), 164.

133 Cady, *From Warism to Pacifism: A Moral Continuum*, 40.

134 Ibid.

135 James T. Johnson, *Just War Tradition and the Restraint of War* (Princeton: Princeton University Press, 1981), 24.

136 Evans, 'In Defence of Just War Theory'.

137 William V. O'Brien, 'The Conduct of Just and Limited War', in *Contemporary Moral Problems*, ed. James E. White (Albany: Thomson-Wadsworth, 2009), 9.

138 Jeff McMahan, *Killing in War* (Oxford: Oxford University Press, 2009).

139 Coady, *Morality and Political Violence*, 64.

Chapter 5
Unmanned Warfare: Technological and Operational Dimensions

Having explored the principles that must be adhered to in resorting to and deploying unmanned force, the focus of this chapter is the technological and operational dimensions of unmanned warfare and, in particular, the dangers of overemphasising what technology alone can achieve in minimising the risk of harm. The task here is to clear the way through some myths and fallacies concerning unmanned systems, which obscure rather than clarify the potential that these systems have to lead us toward a state of affairs involving fewer and less devastating wars. More specifically, this chapter will argue that, while unmanned systems are typically thought to remove people from harm's way and restrict the general destructiveness of war, their capacity to achieve both of those objectives is somewhat limited. Indeed, it will be shown that to some degree, these supposed benefits are part of an erroneous mythology concerning the West's technological dominance and ability to wage harm-reduced and/or casualty-free warfare. In reality, the avoidance or mitigation of risk of harm is often only possible to the extent that such risk can be reallocated or redirected within the ranks of a given military or transferred to various other parties, most concerningly perhaps, private military contractors and noncombatants. This chapter will make use of Walzer's interpretation of just war and further reflect on the moral concerns associated with making delicate tradeoffs between participants in technological war.

Out of Harm's Way?

A review of the existing literature on unmanned systems reveals that much of it promotes the value that such weaponry can have, most notably, 'robots saving lives', 'keeping soldiers out of harm's way' and 'accumulating risk'.[1] In other words, the recurring theme is that of keeping people, both combatants and noncombatants, out of the line of fire. Not surprisingly, considering that military forces have long manipulated the media for propaganda purposes, these themes are commonly found in press releases, government reports and the journalistic media that draws on these sources, but can also be quite readily found in the academic literature:

> Artificially intelligent military robots that *save lives* are preferable to humans (or bombs that kill blindly) ...[2]

> The major rationale for introducing robotic systems is ... most importantly ... to *save lives*.[3]

> ... robots can not only keep soldiers out of *harm's way* and do the most boring tasks, but they can keep going long after a soldier is exhausted.[4]

> Enthusiasm for the use of robots in war stems largely from the fact that deploying robots may help keep human beings out of *harm's way*.[5]

It is the potential lifesaving capacity of these systems that is of note to the military and the public, and understandably so. As was argued in Chapter 3, the existence of a duty to employ unmanned systems can be grounded in the 'military-state contract', according to which it is a moral imperative for states to seek advantage-conferring technologies that protect soldiers and preserve their instrumental and non-instrumental value. The ability to wage war while limiting exposure to harm for many of those involved is, without doubt, a very alluring prospect. While many roboticists and meta-technologists like Arkin claim that fully autonomous unmanned systems are the future of armed conflict, it is yet unclear whether this can be a reality. Indeed, the current technology still requires a sustained human presence on the battlefield, possibly indicating that the risk of physical harm to humans cannot be entirely eliminated. However, Rob Sparrow thinks that there is a subtler problem. He argues that it is quite likely that, as unmanned systems become cheaper to manufacture and operate and thus become more central to warfighting, circumstances will arise in which any existing life-saving relationship between human and machine is reversed.[6] That is, there is a likelihood that distance-orientated unmanned warfare will collapse back into more traditional warfare, where people will need to enter the war zone at close proximity to the enemy, (re)exposing them to harm, and therefore potentially countering any prospective moral gain. In this section and to some extent the next, we will explore a few ways in which this might happen and which should be kept in mind when thinking about the moral value of unmanned systems. Before doing so, it should again be clarified that the concern here is not with the harm possibly generated by a particular physical component of the machine failing, and placing a bystander at direct risk,[7] but rather with how the human-machine relationship that is supposed to keep people out of harm's way, comes to be reversed.

Prior to outlining some circumstances in which the life-saving human-machine relationship is inverted, it must be said that the employment of unmanned systems frequently involves a significantly larger number of people than is often thought. For instance, most unmanned aerial vehicles are teleoperated, though some have supervisory level autonomy. Recall from the scale of autonomy given in Chapter 1 that teleoperated systems are those under direct control of a human being and that those with supervisory autonomy may control certain operations themselves.[8] This means that an unmanned aircraft will have at least one pilot, who may be located within a few hundred kilometres of the aircraft if deployed via a short 'line-of-sight data link'. However, most Western states prefer to wage war from much further abroad utilising satellite uplinks. This means that the pilot will usually work from a ground control station located in a neighbouring country or possibly even on the other side of the world. Admittedly, such pilots are not at significant risk; however, as will be demonstrated in Chapter 7, they are not immune from being attacked and the use of unmanned systems in radically asymmetric conflict may draw hostilities into the suburban areas where the 'remote' pilots are located. Furthermore, each unmanned aerial vehicle deployed in the air requires sensor operators,[9] a crew of surveillance analysts and a maintenance crew. It has been reported that in total, over 150 people are required to keep a current generation Predator drone aloft, while 180 are required for the larger Reaper version.[10] The human resources requirement may improve as the primary flight role of drone operators is automated, allowing them to take on additional roles and functions, but the current ratio of support personnel between manned and unmanned systems is roughly equal.

The use of unmanned systems, even those with satellite command and control links, necessitates that there be people at in-region bases to support their use. Many are of the opinion that US Air Force drones are launched from within the continental United States and flown to Iraq or Afghanistan, but this is not the case. It would involve the rather difficult task of securing the tacit consent of the various sovereign states whose territory would then be overflown. For this reason – coupled with the fact that there is a two-second lag time between the operator's input to the ground control station and the aircraft's response, and the fact that the Air Force appears to take its efficiency imperative seriously by wishing to reduce transit time and maximise operational hours – most unmanned aerial vehicle flights are launched from sites close to their zone of operation. In *Predator: The Remote Control Air War Over Iraq and Afghanistan: A Pilot's Story*, Lieutenant Colonel Matt Martin reveals that there are dedicated, squadron sized, 'launch and recovery elements' that conduct line of sight launch and recovery missions from dangerous places like Balad Air Base in Iraq, which is so frequently under mortar attack that it has been dubbed 'mortarville' by local personnel.[11] The forward-deployed[12] launch and recovery element essentially launches the aircraft and, at a mutually agreed point, transfers control to a crew who is often based in the US, which then goes on to fly the mission. Because of the nature of its functions, the launch and recovery element is always co-located with the aircraft in the potentially harmful forward-deployed environment. Martin also points out that real human beings not only have to go into the war zone to conduct take-off and landing operations for pilots sitting stateside, but also have to train the in-country pilots of those lesser-developed allied militaries with line-of-sight systems.[13] Also, because effective unmanned warfare requires, of course, sound knowledge of the enemy's movements and motivations, this calls for human intelligence operatives to be deployed to the field.

According to Sparrow, a further concern is the need to defend and recover remotely operated unmanned systems that have become inoperable within hostile territory.[14] Because unmanned aerial vehicles are not yet as mechanically robust or manoeuvrable as their manned equivalents and typically lack flare and radar jamming countermeasures, these situations normally arise when the systems are damaged by enemy fire. That said, many systems have also been lost to operator error, inclement weather and component failure. Why is there a need to recover an inanimate machine, given that doing so means placing the lives of those in the recovery team at significant risk?[15] As stated in Chapter 3, an opponent who captures an unmanned system will have nothing but the system itself to show for their effort, and it cannot be duped into divulging confidential information like its human counterpart. However, in the case of unmanned reconnaissance aircraft, it is possible that any sensitive intelligence stored locally could be vulnerable to enemy collection, without the need for any 'duping' and merely a bit of technical know-how. Secondly, as there is often a great disparity in the technological strength of adversaries, the technology itself could be of value to a state wishing to engage in a process of reverse engineering. Information gained through this process could also be of more strategic use in gauging an adversary's capacity to respond to a certain sort of attack. Financial cost is an additional factor. Martin states that when a Predator drone is destroyed, 'the United States loses four million dollars'.[16] All of these concerns came together for the United States when an American RQ-170 Sentinel aircraft crashed in December 2011 after crossing into Iran, and the Iranians subsequently released footage of the top-secret drone to the mainstream media.[17] More often than not, the result in such cases involves deploying helicopters manned with Special Forces troops into a war zone where the enemy may be preparing a deadly ambush.[18] Again, this is quite paradoxical, considering that these weapons are used with the intention of limiting harm and casualties.[19]

While discussion of the psychosocial and political problems of unmanned warfare are reserved for the next chapter, what is also concerning is the way in which the *operators* of some non-autonomous or teleoperated unmanned systems often come to place themselves at personal risk to rescue their damaged robots. Interestingly, some soldiers feel completely disconnected with the systems they operate, while others develop difficult-to-explain bonds. The opening story to Singer's *Wired for War* recounts the story of a military unit that had their unmanned system destroyed by a roadside bomb.[20] In Singer's terms, it sent the unit into a 'deep moral spin', so much so that the commander of the unit wrote a letter home to the manufacturer, as he would a soldier's widow.[21] Later in the book, Singer also tells the story of soldier whose unmanned ground vehicle had been blown up. Upon being offered a replacement system, the sobbing soldier replied, 'I don't want another robot, I want this one. I want Scooby Doo back'.[22] As unexpected as it may be, it seems that the soldier had formed a bond with the machine. Kate Darling, for instance, argues that people are prone to anthropomorphism; that is, that we project our qualities onto certain machines to make them seem somewhat more human-like.[23] There is a range of factors that are said to influence the extent to which we are likely to anthropomorphise a robot and potentially put our life at risk in retrieving such robots,[24] but Bartneck and Hue have demonstrated that there is a relationship between the perceived intelligence of the robot and concerns about its destruction. In this case, the value of the system is derived from the machine qua human-like entity. Ian Roderick, on the other hand, perhaps more plausibly argues that it is actually about 'robot fetishism' and that the cases above need to be contextualised as a sign of social valuation and mediation.[25] Under this sort of account, the value is derived from the machine qua itself, insofar as the machine is capable of taking on labour classified in the technological age as being 'too risky' to be extracted from humans. The point here is that soldiers sometimes go to great lengths to rescue unmanned systems, because they so highly value them as human surrogates in what is essentially a process of risk transference (whereby risk is shifted from one party to another/others). When it comes to humans rescuing machines, however, the process has gone somewhat awry.

Whether unmanned systems are rescued because of some human-machine bond, or in recognition of their specific value in accumulating or reducing risk that would otherwise burden human warfighters, the more significant moral matter concerns the risk to humans once these systems have been recovered. The problems encountered with unmanned systems can be quite wide ranging. Historically, most widely used military technology has been 'ruggedised', or otherwise designed so that parts can easily be replaced or repaired in situ by regular personnel to ensure it can be placed back in operational service with little downtime.[26] Many unmanned systems, namely basic ground vehicles, are also designed with such considerations in mind.[27] However, the introduction of technologically advanced systems with complicated weapons, sensors and communications equipment has meant that military forces have had to establish so-called 'robot field hospitals', in which humans are put at risk to service robots.

During the latest conflict in Iraq, the Joint Robotics Repair Facility located at the United States Military's main base near Baghdad International Airport, repaired as many as 400 robotic systems per week.[28] Because the military struggles to attract and retain appropriately qualified personnel, and because cost-saving initiatives have focused on converting as many military billets as possible to less expensive civilian positions, many such facilities are now staffed partly or completely by non-uniformed human personnel.[29] These non-uniformed personnel are usually contractors working for large robotics and aerospace firms, lured into such risky roles by the 'hazard pay' or 'danger money' that accompanies them. With the growing sophistication of the technologies of war, including unmanned systems technologies, it is likely that we will see military forces ever more frequently trying to take advantage of civilian expertise. These civilians, who act under financial coercion, are not, however, bound by the same legal or moral standards as military personnel, and this presents some obvious problems in such a rule-governed activity. It should additionally be noted that moving civilians into the line of fire, which is troubling in itself, places an additional burden on the soldiers who are tasked with their protection. Thus, unmanned systems do not strictly speaking keep people out of harm's way, and this presents a real challenge for those who seek to justify their use on this basis.

Drone Complacency, Dependency and Servitude

The moral rationale for employing unmanned systems could also be seriously undermined by a multitude of other situations. In this section, the focus is on three cognitive phenomena that can reduce the efficacy of unmanned systems and present a range of strategic problems of moral importance, and that therefore deserve greater attention than they are usually given in the literature: namely, complacency, dependency and servitude.

We begin by examining complacency at the individual level. Sparrow writes that as soon as soldiers recognise that unmanned systems can play an effective role in conducting military operations, they will begin to trust[30] them to carry out routine tasks that they would otherwise have had to perform themselves.[31] Les Hauck also writes that while the Wright brothers were far from lackadaisical in carrying out their first flights, today, with innovations that allow aircraft to fly for extended periods of time without any input, it is much easier for pilots to place faith in their systems.[32] It seems that, whenever people use a new form of technology, including those partly automated instruments used in the process of warfighting, and they use it repeatedly without fault, they tend to forget about the potential for failure.[33] Of course, the extent to which people over- (or under-) rely on unmanned systems will also depend on how many tasks they are given to perform.[34] To illustrate the problem, take the case of an explosive ordnance disposal team tasked with the daily clearing of a well-travelled road. The team uses an unmanned ground vehicle to check the path and soldiers follow behind, performing a precautionary secondary sweep. The robots have always been effective and, after two months of clearing the same road, the team's platoon sergeant decides that the secondary sweep is largely unnecessary, so it is eliminated every other day. On one of the 'other days', the robot, with its slight margin for error, misses an improvised explosive device that destroys a passing convoy. Had the road been swept manually, the explosive would have been detected. This hypothetical case suggests that the complacency, which can accompany the use of unmanned systems, puts human life at risk, and this potential for harm is exactly what the machine is supposed to help avoid. Unfortunately, complacency is difficult to combat when the same action is being repeatedly performed, which is often the case in extended military operations.[35] This is only compounded by the fact that many military forces tend to introduce new weapons systems without altering doctrine or documented best practice.[36]

Also of interest and importance is the relationship between technologically induced complacency and technological dependency at the military or state level. As with individuals, some military forces have tended to rely on unmanned systems. This is potentially dangerous for several reasons. The first is that many unmanned technologies were designed to detect large and fairly conventional forces, which do not necessarily reflect the character of modern warfare.[37] Another relevant factor is that non-conventional targets have adapted to fighting in a theatre of war that is inundated with sophisticated weapons, such as drones that can sense and detect individuals on the ground. As one Canadian task force commander put it, 'all the stupid Taliban are already dead'.[38] The intelligent terrorists exploit cover or camouflage that can be found in most 'theatres of war'. For instance: cave systems, bunkers and urban areas provide overhead cover and some background clutter; foliage degrades the effectiveness of most sensors, even the infrared versions employed on modern unmanned aerial vehicles such as the Predator.[39] Finally, some unmanned systems have weapons that can target a particular kind of enemy technology, but this is only effective when the enemy employs that particular technology. This does not suggest that unmanned systems and their modern sensors are without benefit. Even the most well organised armed group will find it difficult to cover territory without coming into range of the enemy's sensors at some stage. The argument here is that, at the moment, it would be imprudent and perhaps morally reckless, for a state or its military force to rely too heavily on unmanned systems technologies to defeat an adversary that has learned to exploit its surroundings and their enemy's weaknesses.

Sparrow further remarks that states and their military forces, which can be complacent about the capabilities of unmanned systems and depend on them too much, can be dangerously wrong about certain conflicts. Indeed, conflicts may be initiated with the view that they can be won without placing warfighters in harm's way,[40] only to find at a later stage that victory – or the abortion of the operation – can only be achieved by committing more conventional forces, requiring the involvement of human beings in the zone of hostile operations.[41] This could occur, he says, either because the system(s) malfunctioned, or because changing circumstances result in the operation being beyond the reach or capability of the system(s) in question, or because the operation was ill conceived or impractical to begin with.[42] History is replete with examples demonstrating the moral cost of fighting wars to which states are unwilling to commit the necessary resources for victory. The American record, in particular, is mixed, at best. Many consider the Vietnam War to be proof that technologically induced complacency and dependency can lead to disapproval on the home front and, ultimately, to failure. Critics wrote of 'blind technological fanaticism, hubris, and overconfidence as the United States attempted to fight a remote, antiseptic war'.[43] The strategy in Vietnam was one of airpower dominance, but America underestimated just how ineffective its technologically sophisticated fighter jets would be over a land covered in dense jungle foliage,[44] and the extent to which such technology would antagonise the civilian population, without whose political support victory was not to be possible. Iraq and Afghanistan are perhaps more recent examples of the resounding effects of technological complacency and dependency. In Lawrence J. Korb's words, the idea that war could be won with the effort of 'three guys and a satellite phone', proved to be quite dangerous.[45] Put simply, their application of the probability of success principle was clouded and they could not succeed without the application of more robust military force.

Another problem with technological dependency is that it may degrade a military force's ability to mount a robust defence against internal or external violence or harm, which, as outlined in Chapter 4, is its primary function and obligation under the contract that governs its relationship with the state and its people. Firstly, and perhaps most seriously, if unmanned systems continue to be deployed quite widely because they are *thought* to perform better than humans, reduce costs and so on, it might be that a gradual loss of skills or knowledge occurs over time.[46] That is, by depending on technology that promotes the waging of distance warfare and the removal of human warfighters from the battlespace, military forces may actually be contributing to a loss of their conventional warfighting skills, which may have to be relied upon if hostilities arise that cannot be overcome with the use of technology alone. After all, unmanned systems operators do not need physical strength or combat experience, nor do they need to possess the virtue of bravery.[47] Take drone operators. A comprehensive study conducted by the Ohio School of Aviation Medicine revealed that currently, the primary accession source is rated pilots from manned airframes who have been selected to 'cross train' for unmanned systems service.[48] These pilots are typically of the rank of Captain or higher and have a significant amount of flying experience. They serve on a drone assignment for three to four years.[49] Over these years, these experienced pilots may come to be deprived of any real world flying experience or training. Similar statements can be made about soldiers and seamen and the possible degradation of their skills. The concern here is that, if unmanned systems operators return to direct contact with the enemy, they may be susceptible to being harmed more easily and be less effective in fulfilling their obligations to the state and its citizens. This means that, in such cases, the risk of harm is transferred in such a way that it completes a full circle: from human to machine and back to the human. This, of course, is not the desired effect that unmanned systems are intended to have.

Patrick Lin, George Bekey and Keith Abney argue that this degradation argument does not have much force as a general objection to unmanned systems, because the benefits of employing the technology in question still seem to outweigh any potential loss.[50] They point out that, for instance, while our ability to perform mental arithmetic may have suffered somewhat following the invention of the calculator, computer and spreadsheet, we would rather keep these tools even at the expense of being able to perform complex mathematical calculations in our heads.[51] By way of response, we must be careful when reducing the matter to a simple cost-benefit analysis, so that it becomes too narrow in scope. Taken alone, it might be granted that we are willing to keep the technology at the expense of the degradation of our skills. However, this problem is only one of many that will later be put into the context of the process of risk transference. It is certainly possible that, if we come to rely on robots in waging most of our conflicts, we may eventually have so few conventionally trained warfighters with genuine combat experience that it becomes impossible to successfully achieve the goals we have for waging warfare. In fact, in 2008, the US Air Force began drawing drone pilots from ranks with no 'real' flying experience whatsoever.[52] It may be that we are witnessing the gradual death of the traditional combat-trained warfighter, potentially presenting force readiness problems at a great cost to the state and those whom the armed forces are charged with protecting. Employing private military contractors on short-term agreements might be the only solution to fill the gap in the force structure in a limited timeframe, should conflict arise. This is problematic not only because private contractors are not bound by the same moral standards, but also because they may not be as well trained as conventional standing armed forces.

Additionally, we have to think about our moral responsibilities and the transference of risk in the context of the displacement of military personnel themselves. As Arkin states, much of the underlying premise for the use of military robotics is the improvement of working conditions and the elimination of the three 'D' jobs: dull, dirty and dangerous.[53] The standard argument is that displaced workers go on to find more meaningful work.[54] However, this does not necessarily hold with respect to warfighters, who may have spent their whole lives – social, physical and psychological – preparing or maintaining readiness to defend their nation and its people. For service personnel, there are few 'higher-value' jobs than serving one's nation through rendering active military service on the battlefield, so this assurance of finding alternative and meaningful work in a more passive role is likely to be of little consolation to a displaced soldier. Many military personnel are offered and accept roles working with the robotics that displaced them.[55] Others unequivocally reject such roles (they are often career fighter pilots who have been pushed to extremes through their training and service, and who value the human element in warfare). After all, a life operating or repairing unmanned systems involves an entirely different – and many would argue, much less fulfilling – form of military service. Thus, in another role reversal, unmanned systems may save the lives of some, but they also create lives of misery and unfulfilling service for others.

New Weapons, New Tests, New Targets

One remaining problem with the use of unmanned systems concerns their potential to induce conflict or cause suffering and harm, distinct from the more psychosocial and political aspects of unmanned systems use, which are considered in the next chapter. It needs to be recognised that unmanned systems, whilst having a long evolutionary history, are still first or early second-generation weapons. It was not until fairly recently that these systems came to be utilised so widely in military operations: the large majority of systems having been hurriedly put into use since the beginning of Operation Iraqi Freedom in 2003.[56] As with any technology rushed into service, errors inevitably occur and these need to be corrected in the redesign phase. With regular software and technology, it is a relatively simple matter of applying a 'patch' or installing a component update. The problem, as Lin explains, is that with military systems, the moral stakes are significantly higher, since the role of these systems is to save lives, usually through taking those of others.[57] Therefore, programming issues or other errors with unmanned systems will often be fatal, as was the case with a semi-autonomous robotic cannon that malfunctioned when used by the South African National Defence Force.[58] During a shooting exercise with live ammunition, the anti-aircraft weapon fired uncontrollably into a crowd of spectators. Because this Swiss-made cannon can automatically lock on to targets and reload when its magazine is empty, nine soldiers died and 14 others were wounded in this one incident alone, despite the efforts of an artillery officer who risked her life to stop it.[59] We also know that if applications for unmanned systems technologies exist, manufacturers will continue to build them and the technology will no doubt improve over time. However, as Lin and his colleagues write, it would be 'naïve to believe that another accident with military robotics will not happen again'.[60] Thus it seems that both morality and prudence call for greater testing of unmanned systems before they are acquired, and even more rigorous testing before they are deployed in any large-scale operations which have the potential to impact the lives of actual human beings.

Today's unmanned systems are already tested in controlled laboratory environments, but no manufacturer can say with any certainty that a particular system is error free and will perform as per the design brief. As noted by Colin Gray, it is notoriously difficult to determine the effectiveness of a new weapon to warfare in peacetime.[61] With unmanned systems, the manufacturing process is often also a collaborative effort and involves many different parts and potentially millions of lines of code.[62] This means that no one individual can certify that every element of the system will function without fault in the battlespace. Even field-testing can fail to uncover some errors of potential concern. Just as it is not possible to run a complex piece of commercial computer software through every possible end-user scenario, it is not possible to put a military robot through everything it could encounter on the battlefield.[63] States that have come to rely on unproven technology therefore face the challenge of demonstrating to stakeholders that their military forces' expenditure on new weapons systems is justified. Vladimir Slipchenko, an influential Russian General and military scientist, has gone so far as to suggest that past wars were little more than tests aimed at justifying weapons acquisitions.[64] It should be observed, he says, that:

> The Pentagon buys from the military-industrial complex only those weapons that have been tested in conditions of real warfare and received a certificate of quality on the battlefield. After a series of live experiments – the wars in Iraq, Yugoslavia, and Afghanistan – many corporations in the US military-industrial complex have been granted the right to sell their precision weapons to the Pentagon.[65]

While a controversial claim, Slipchenko is right to warn us of the dangers of the 'military-industrial complex' and its war-mongering ways, as was Dwight Eisenhower, who coined the term in his presidential farewell speech of 1961.[66] Eisenhower reminded his audience that the immense military establishment and large arms industry for which the US is so well known, is actually a fairly recent construct. American manufacturers have long held the capacity to make ploughshares and elementary weapons, but it is only with the latest of our World Wars that the US has felt the need to develop and maintain an armaments industry of such vast proportions.[67] Medea Benjamin writes that while the United States is in the midst of recession and the once prosperous manufacturing towns (such as Flint and Detroit in Michigan) are now little more than recruiting grounds for military enlisters who prey on the unemployed, the large corporations and defence contractors which manufacture the modern tools of warfare continue relatively unscathed.[68] They are ever ready to sell war just as pharmaceutical giants sell sickness.[69] She goes on to say that the Pentagon's 'war chest' for drones exceeds five billion dollars and that global spending on unmanned systems is expected to total more than 90 billion dollars over the next eight years, with American firms benefitting more than those of other nations.[70] General Atomics, a San Diego-based private company manufacturing unmanned aerial vehicles (such as the Predator and the Reaper), sold nearly three billion dollars' worth of equipment to the military over the last 10 years.[71] Corporations such as Raytheon, who provide the software that allow the military to get real-time intelligence from unmanned systems, as well as Boeing and Lockheed Martin, both well-known aerospace firms with vested interests in unmanned systems and their munitions, have annual revenues of over 100 billion dollars and spend almost 15 million dollars per year on lobbying.[72] In early 2011, a 'Congressional Unmanned Systems Caucus' was also established, which advocates an industry that replenishes its political 'war chest'.[73] This industry is constantly developing new systems that must also be tested and, as Slipchenko writes, when the US military-industrial complex demands test-bed wars [platforms for large scale experimentation] from its country's political leadership … it gets them'.[74] The involvement of the above-mentioned corporations in past wars, such as the 1991 Persian Gulf War in which the Patriot missile was tested, partly validates his claim.

Slipchenko explicitly hypothesised that, before combat operations in Iraq even began, America's *primary* objective would be to test its latest precision weaponry and that in the initial stages, this conflict would not involve a ground war in the classic sense.[75] Of course, his prediction of a non-classic ground war was later proved right. On the evening of 21 March 2003, the United States began operation 'Shock and Awe', in which they gained rapid military dominance without large numbers of ground troops. The world saw combined allied forces launch over three thousand laser-guided bombs and cruise missiles, all of which fell on key Iraqi targets over a period of only two days.[76] Upward of 700 precision air attack sorties were flown against Iraqi ground forces, rendering them largely unable to defend Baghdad.[77] The so-called 'coalition of the willing' worked in accordance with its 'maximin' principle:[78] achieve maximum effect with minimum force. While only a handful of drone missions were flown in the initial invasion, these systems have since been used extensively in the hope of fulfilling the coalition's maximin aim. Even with this in mind, to suggest that testing weapons was America's primary objective in resorting to war is undoubtedly too strong a claim. However, it seems entirely plausible to suggest that such testing was factored into the *jus ad bellum* calculus.[79] It can reasonably be said that the development of unmanned weaponry might provide secondary or unjust cause for waging warfare, raising concern for any application of the right intention principle. Remember that this just war principle serves to limit war by ensuring that it is fought primarily for a just cause and that less 'pure' (although not abhorrent) intentions are factored in only on the condition that they do not dominate the just cause. The problem is that weapons testing may be creeping up on the scale of importance and become more dominant than it should. This means that there is a greater likelihood of states applying unmanned systems or, at worst, being lead into wars that they would not have otherwise engaged in or waged. The moral consequence is that this places everyone at greater risk of harm than expected and justified.

Another politico-technical factor that might be responsible for a greater incidence of warfare involves sovereignty or, more particularly, how unmanned systems enable war to be waged without invading sovereignty in the traditional sense. More often than not, drone strikes and the like are spoken of and written about as an intrusion on the sovereignty of states, e.g. Afghanistan, Pakistan, Yemen and Somalia, where drones have very publically been used. For instance, the President of Pakistan, Asif Ali Zardari, has publically stated that he 'cannot condone violations of [Pakistan's] sovereignty even when they are done by allies and friends. [He] would much prefer that the US ... allow [Pakistan] to take care of this problem on [its] own'.[80] Under a strict interpretation of international law, particularly Article 2(4) of the United Nations Charter ('members shall refrain in their international relations from the threat or use of force against the territorial integrity or political independence of any state'), unmanned aerial vehicles would also seem to infringe on these states' sovereignty. After all, these systems frequently cross international borders and use lethal force to eliminate targets. However, a counter-case might be made that unmanned systems actually placate concerns over sovereignty and facilitate military operations in sovereign territory. For example, a British policy paper on unmanned systems suggests that the extensive operations that the United States and its allies have recently conducted in Pakistan and Yemen would not have been possible, were it not for the existence of these systems.[81] Neither Pakistan nor Yemen would have let manned aircraft into their territory, just as they would not allow large numbers of combat troops to pass through their territory to conduct an operation they do not approve of. Unmanned systems, which some say are no more invasive than satellites in orbit,[82] allow sovereign states to convince others that unmanned systems do not pose a threat to their sovereignty like other military activity. In such cases, it seems that the use of force and its probability of success is entirely a function of the technological capacity to wage unmanned warfare. Unlike satellites, however, these systems are often weaponised and their potential failure, when invading and operating in a sovereign territory, can put the citizens of these sovereign states at risk.

Importantly, the development of unmanned systems opens up not only new territories and conflicts to states and their military forces, but also the possibility that new targets may become available for attack within these territories and conflicts.[83] In Sparrow's words, insofar as they make possible more discriminate and proportionate uses of lethal force, unmanned systems risk allowing their operators to attack targets that they would not otherwise have been able to attack under just war theory, without causing a great deal of collateral damage to surrounding areas and risking an excessive number of casualties.[84] In their article on drones in contemporary conflict, Sarah Kreps and John Kaag cite Air Force Lieutenant, General Buster C. Glosson, who attests to the fact that these weapons open up new targets.[85] Glosson writes that in order to shut down industry in World War II, it was a matter of targeting entire complexes because of the inaccuracy of the available weapons.[86] More often than not, this involved dropping a large number of bombs and hoping that one would hit the desired target, a view of which was often obscured by cloud or smoke. Glosson continues by saying that today, it is possible to strategically hit only a couple of key buildings.[87] The argument is that, with weapons such as unmanned systems, what historically could only be accomplished with volume can now be achieved with precision. Of course, this means that military forces may be tempted to attack more targets, namely those in congested areas that were previously accessible because of the imprecision of earlier weapons. The obvious reaction consists in suggesting that if these are legitimate targets in a just war, this ought to be no problem. However, an accepted fact of just war theory is that military conflicts produce civilian casualties and, if the number of conflicts rises, so too will the total number of casualties. This remains a horrible thing despite the fact that civilian casualties are lost in pursuit of a just cause. Of course, it may be that if unmanned systems massively minimise collateral damage and casualties, and become increasingly better at doing so, then the number of deaths will decrease or remain relatively steady compared to earlier situations. However, for technological reasons explored more thoroughly in the next section, military forces may only be able to achieve greater utility with unmanned systems at a cost to the civilian population, further challenging the justification for employing unmanned weaponry.

Urban Warfare and Technical Limitations

As the previous discussion has suggested, if unmanned systems improve precision, they might leave a range of previously inaccessible targets vulnerable to attack. In this section, we examine the implications of this with reference to the concern that many of these targets are likely to be located in urban centres and that, while unmanned systems may have improved precision targeting capability to the point that it is now possible to destroy a single specific building in a complex rather than the whole complex, these socio-technical systems are not yet capable of fulfilling the more exacting moral demands which come with urban warfare and counterinsurgency operations. Moreover, it will be argued that while unmanned systems technologies may improve precision to some degree, they exacerbate rather than relieve the problem of exposing noncombatants to higher levels of potentially lethal risk, countervailing the duty to employ these systems.

The argument put forward at the end of the last section was that insofar as unmanned systems improve precision, we may see a shift away from the imprecise bombing campaigns that typified military operations in World War II, toward what might be more efficient, effective, discriminate and proportionate applications of force, all with the aim of overcoming acts of aggression and the collateral losses of past wars. Colonel Phillip Meilinger, commander of the US Air Force's School of Advanced Airpower Studies, believes that we are already seeing the effects of the technologically induced shift toward 'airshaft accuracy'.[88] He states that currently 'there is no logical reason why bullets or bombs should be wasted on empty air or dirt. Ideally, every shot fired should find its mark'.[89] Today's targets are no longer massive industrial complexes or sprawling military installations, nor are today's weapons the size of the B-52 bombers that would, in the past, have been used to drop bombs on such large targets. Both weapons and their targets are shrinking. This raises the following problem: political and military decision makers may become *too* comfortable with using unmanned systems in attacks deep in major cities, towns and villages. In addition to highlighting the risk of increased rather than decreased harm, this also potentially constitutes another example of the manifestation of complacency. In *Virtual War: Kosovo and Beyond*, Michael Ignatieff suggests that this may already be a reality.[90] Having visited the ruins of a Serbian TV transmission centre attacked during the North Atlantic Treaty Organization's bombing of Belgrade in 1999, he writes that:

> the destruction was awesome because it was so unbelievably precise: not the administration block next door, not the still intact studios under blue reflecting glass on the other side, but just this one single studio area, methodically crushed, as if a hand from the sky had pushed down floor by floor, seizing hold of human beings, chairs, tables, equipment, monitors and plunging them into the pile of debris at the bottom.[91]

It is certainly good that military forces need not risk the broad-area bombings characteristic of earlier wars; nevertheless, this quote reflects that precision gives military forces a greater freedom to carry out attacks closer to noncombatant-manned areas (both in the enemy's territory and enemy-occupied territory) than ever before in military history. This is concerning, for while Ignatieff reports of a single precision strike, he fails to report the destruction caused by another in the same attack. This other attack, which is arguably more indicative of the effectiveness of 'precision strikes', left a large detonation crater between two homes, injuring the residents, destroying the road and severing the water supply.[92]

Unmanned systems and the improved precision that they promise essentially bring warfare into the streets of our cities, towns and villages. This urban environment comes with a unique set of challenges that have only recently started to receive attention in relation to drone usage. To understand why urban operations deserve special attention and what makes them operationally and morally challenging, we have to think about the characteristics of this man-made environment. The three important factors for this purpose are: 1) the physical features; 2) the infrastructure; and 3) the population. All of these characteristics are intertwined and, in some respects, inseparable. To begin with, take the physical characteristics of a typical large city. As Alice Hills writes, these city spaces, with their large buildings, towers and often-sprawling complement of residential homes, represent a complex 'multidimensional blend of horizontal, vertical, interior and exterior forms, superimposed on natural relief'.[93] If weapons are to be used in cities, and to a lesser extent, towns and villages, they must then be capable of multidimensional manoeuvring around buildings and through often-vast networks of underground tunnel systems. Moreover, Hills notes that urban terrain provides cover and concealment to the enemy and limits observation distances and engagement ranges.[94] This limits the effectiveness of unmanned systems and gives rise to a number of delicate moral tradeoffs. A Predator pilot, for instance, has a certain range at which missiles must be launched in order to be effective. If the urban terrain limits this, the operator will be forced to choose between not firing the weapon at all, or going ahead whilst knowing that the weapon may not be as discriminate as just war theory demands in these particular circumstances. Of course, if an operator chooses the latter course of action, the urban terrain may limit the line of sight and any ability the operator has to abort because of communications interference. The operator's ability to comply with just war theory is effectively restricted.

Another problem consists in the fact that, despite just war theory requiring military forces to forewarn a civilian population of plans to attack their city whenever reasonably practicable, an urban area worthy of being seized is rarely going to be deserted. Securing a city or simply conducting warfare in such an environment will therefore involve operating among and controlling its population. While harm caused to noncombatants is a common problem, casualties in urban warfare in particular have historically been high, and so too have been the levels of brutality.[95] The United States, Britain and other allied nations that employ drones emphasise the care they take in their targeting and review processes so as to minimise the harm caused to the civilian population by unmanned warfare, but care alone cannot overcome the human element that comes with operating in urban environments. Sensor and targeting technologies may be good enough to destroy static buildings with incredible precision, but it is much more difficult for operators to distinguish legitimate human targets and objects from the non-legitimate targets and objects surrounding them. Combatants have also come to realise that the battlefield is increasingly integrated with population centres and have purposely started intermingling with noncombatants,[96] exploiting the fact that scrupulous states will be reluctant to target them, just as they would be reluctant to target an arms depot located next to a Red Cross hospital. Moreover, many combatant groups either lack uniforms (as with many armed insurgent groups) or have consciously decided not to wear them. This is concerning because uniforms function not only to limit friendly fire incidents, but also to draw enemy fire to those in uniform and away from those whom are not taking part in the hostilities.

Further muddying the distinction between combatant and noncombatant are private military contractors, as referred to earlier. The question here is not whether private military contractors should be afforded combatant status – this categorisation has been extensively discussed elsewhere[97] and is not one that can be deliberated on here – but whether their presence on the battlefield exacerbates the difficulties associated with discriminating between combatants and noncombatants via unmanned system. Private military contractors currently employed in the Middle East do not usually provide direct, tactical military assistance of the sort that one would normally get from experienced military personnel. Rather, these 'civilian contractors' are primarily given roles in logistics support, the training of the new Iraqi military and police force as well as various guard duties.[98] These are still dangerous roles and, while they are not hired explicitly to wage warfare, contractors providing military security services (such as installation protection and convoy escort) have often been forced into direct combat with enemy forces. The governments of Angola and Sierra Leone, for example, have employed private military contractors explicitly for the purpose of stopping forces that their own troops could not.[99] The future may also see them operating drones. Since it can already be notoriously difficult to discriminate between combatants and noncombatants, it could be argued that the protection of the civilian population is indirectly affected and placed at greater risk through the blurring of boundaries, which comes as a result of the increased use of private military contractors. New non-state actors such as these only sow confusion with respect to the moral principles with which unmanned warfare must comply. Rebecca Johnson argues that, since these new non-state actors challenge the status-based approach to combatancy in part by their intermingling with the civilian population, the test of material participation – used to determine whether one is actively involved in a conflict –

becomes much more central to understanding when and under what conditions it is permissible to target individuals from afar.[100] That is, more emphasis is placed on monitoring an individual's conduct in war and how this contributes to the hostilities, prior to making any lethal targeting decision.

It is in this regard that proponents say unmanned systems play an important role. Johnson goes on to argue, for instance, that drones play an increasingly important role in identifying membership of armed groups through an improved capacity to identify material participation.[101] The intelligence, surveillance and reconnaissance capability of unmanned aircraft, which can loiter for extended periods of time, can supposedly improve an operator's ability to monitor an individual's movements for days at a time in order to determine individual instances of material participation or actual membership in a particular armed group. What is more, she says, current technology allows for individuals to be followed by a drone until such time as the individual is away from noncombatants who may be at risk.[102] But, as was already said, combatants will often stay embedded in the urban area until surveillance has ceased. One might expect that, as sensor technologies continue to improve, the quality of the targeting decisions made by operators – up until the point when such decisions are fully automated – will also improve and lead to more discriminate and proportionate warfare. In response, it must be acknowledged that sensors will improve, and so will *some* of the decisions that are based on the data they collect, but it is not as clear whether we are likely to see any further significant improvement in our capacity to distinguish legitimate targets from illegitimate targets in the near future. As Michael O'Hanlon illustrated in *Technological Change and the Future of War*,[103] technological improvements are likely to come in the form of smaller and perhaps cheaper sensors, which will lead to more sensors possibly being placed on more unmanned platforms, thus increasing the overall quantity of information available. For example, a new video capture technology, which is under development by the United States military, dubbed the Gorgon Stare – named after the Greek mythological figure whose gaze turned enemies into stone – will reportedly increase the number of video feeds from a single Predator drone from a couple to over 60. This constitutes a vast increase in the amount of data collected and this would seem likely to improve discrimination and therefore represent a moral improvement.[104]

What will be significantly less quick to improve, however, is the quality of the information and the operators' ability to process it. Kreps and Kaag question the usefulness of all of this information and argue that these weapons still do not ensure that a selected target is a legitimate one. That determination, they argue, 'is of a legal and ethical, rather than technological, nature … Technology, even when it is sophisticated and precise, cannot determine the proportionality of strikes … In short, technology does not provide an easy way to comply with [the] … principles of distinction and proportionality'.[105] Johnson counters that this is an obvious and morally insignificant point, since the larger question surrounding combatancy is, by definition, a human problem. She argues that so long as there is a human operator in the decision chain, there is no significant moral difference between Reaper drone attacks and those of manned Strike Eagle fighter jets, except that one might allow for greater precision.[106] However, Johnson seems to misunderstand the argument. Kreps and Kaag acknowledge that the larger moral question concerning combatancy does not arise as a result of unmanned systems. They simply seek to acknowledge that unmanned systems facilitate it and that there may be problems involved. Their point, if understood correctly, is that unmanned systems technologies make the goals of distinction and proportionality more ambiguous and their achievement more problematic. As Johnson has argued, advocates often frame the technological mediation vital to what we call the 'drone stare', as a morally unproblematic ability to see the truth of a particular situation, or achieve a totalising view of the target or situation in question with which to make targeting decisions.[107] The reality, however, is that it is far from clear whether the sensor systems of unmanned platforms are capable of relaying the moral reality of a situation on the ground to a sensor operator who might be sitting half a world away. Rob Sparrow, for instance, argues that while sensors allow operators to visually discern tanks from pickup trucks, combatants from noncombatants, and hospitals from military installations, it may be that there are some aspects, which are essential to making ethical targeting decisions, but which are so subtle or embodied that they simply cannot be conveyed by even the best sensors.[108] These could be, for example, certain actions, experiences or values.

In practice, there is no way to judge the significance of this problem. As Enemark writes, this is due to a dearth of reliable empirical data on combatant and noncombatant casualties resulting from unmanned warfare and precision drone strikes.[109] US officials have claimed that while the frequency of drone attacks has been increasing, the number of civilian casualties has actually declined. A *Foreign Affairs* article published in mid 2011, for instance, cites government officials as stating that in Pakistan, over the previous two years, there were 'fewer than thirty civilian deaths' caused by America's drone platforms.[110] In a slightly earlier article in *Foreign Policy*, the same authors reference another official who claimed that 'just over twenty' civilians had been killed in the prior two years while 'more than four hundred' combatants had been eliminated.[111] Similarly, a 2010 BBC story reported that over 650 warriors had been killed compared to 20 civilians: a combatant to noncombatant 'kill ratio' of roughly 30:1.[112] Many, however, suggest that unmanned systems technologies do indeed hinder their operators' ability to conduct discriminate and proportionate attacks and that the brunt of the miscalculation, and the distribution of death, falls much more heavily on noncombatants. Avery Plaw has compiled information from three independent drone accuracy databases and indicated that, while the figures are improving, the kill ratio might be as poor as two to one – a figure that points to a shift in the burden of risk that is unlikely to be regarded as morally tolerable.[113]

Unmanned Warfare and the Risk Economy

Before we can begin to think about what level of risk transference, if any, is tolerable under just war theory, we have to consider how and why the risk-transfer economy (or system) has become central to the Western way of war. Part of the reason that risk mitigation has become so important can be attributed to what Ulrich Beck calls a 'technological moralisation', which removes the need for soldiers and statesmen to think critically and abstractly about ethical or moral imperatives. Beck states that this sort of risk calculus exemplifies a form of 'ethics without morality'.[114] In the technological age, it has the unfortunate consequence of being seen by some as reducing morality to a mathematical equation. As Christopher Coker writes, we often find ethical demands being examined in terms of technical data such as mortality rates, with whatever puts the human body at risk being considered unethical.[115] He says that when we invoke the language of risk, we invoke an amoral language or a moral language that involves an inherent technological bias.[116] Yet Coker arguably underestimates the operationalising power and moral importance of risk in war. On a contractual basis, war is after all an agreement to resolve conflict through recourse to arms, even the most basic of which are technological artefacts. The problem, according to Coker and as discussed in this chapter, is that in modern warfare we are quantifying dangers, converting them into risks, and seeking to reduce or eliminate them at a greater rate than ever before.[117]

Despite the modern obsession with identifying, tallying and responding to risk, it is not always possible to eliminate or compensate for the uncertainty that underpins it. As a consequence of this, some states have naively attempted to deal with this larger problem and reduce the negative effect or probability of risk by seeking to shift the burden of risk onto others. As Martin Shaw writes in *The New Western Way of War: Risk-Transfer War and its Crisis in Iraq*, modern war involves the comprehensive transfer of risks away from Western states.[118] It often involves multiple transfers: some involve transferring a particular risk from the West's combatants to other enemy combatants and noncombatant civilians; others are more systematic transfers. Shaw says that, in the first instance, the physical risks of war for Western states are transferred from governments to their own militaries.[119] There is nothing unique or unusual in this though. As we saw in Chapter 3, it is democratically elected governments that outsource their protective function to military forces and uphold the military-state contract, which establishes their moral obligation to ensure that they are effective and efficient in that function. As already explained, these efforts have a crucial bearing on the employment of unmanned systems and the way risks are subsequently transferred from military personnel to these systems. In utilising these systems in warfare, the West seeks to impose the majority of life-risks on enemy combatants, and to minimise those for civilians whenever possible. In doing so, to use Shaw's words, they try to 'transfer the burden of risk from civilians to where it belongs, with the armed enemy'.[120] However, this has not eventuated in practice. As evidenced by the preceding sections, the West is effectively reallocating risk amongst combatants to private military contractors and even noncombatants.

In considering whether this is permissible and how unmanned systems ought to employed, if at all, we have to ask ourselves what level of risk these groups are required to assume and whether we can prioritise any of these groups when it comes to engaging in a process of risk management. What degree of risk transference, if any, is tolerable within and among them? Put another way still: is it morally justifiable to increase the risk to one or more of these groups (or the persons within them) in order to reduce it for another? First of all, there are some questions that must be raised about transferring risk between combatants of a particular military force. As we saw in Chapter 3, soldiers are typically motivated to enlist by an induced sense of patriotic duty created under the social contract and, prior to serving, take a general oath to serve and defend the state, its leaders and citizens. They are willing to risk their lives in defence of their families and way of life.[121] Members of the armed services therefore receive general combat training so that, in the event of a total war, all military personnel have a basic level of combat readiness. That said, when someone joins the all-volunteer armed forces of a democratic state, they are usually given a preference as to what they want to specialise in and can choose from a range of options that are more or less risky in nature. It could therefore be argued that, if unmanned systems end up requesting of soldiers, who have selected not to be involved in the use of these systems, to put their lives at risk in their defence or to complete a mission they failed to achieve, this represents a change in the contractual relationship between soldier and state, rendering the risk-transfer unethical.[122] There is also the question of transferring risk between the combatants of the warring parties, but this discussion is left for Chapter 7 and a critique of radically asymmetric unmanned conflict.

If this sort of risk-transfer war raises questions concerning the morality of targeting combatants with unmanned platforms, those concerning noncombatants go straight to the heart of just war theory and, in particular, *jus in bello*. We know from the requirement of discrimination that parties to any given conflict must make a concerted effort to distinguish between combatants and civilian noncombatants and that military operations should be directed only toward the former. Armed force ought only be deliberately directed at military targets and rarely, if ever, at civilian targets. However, as briefly remarked in Chapter 4, many theorists stipulate that the principle of discrimination involves a more morally demanding obligation to act with 'due care'. Walzer, in particular, argues that not to intend the death of civilians is insufficient. In his words, warriors – a class in which he would presumably include unmanned systems operators – must display a 'positive commitment to save civilian lives'.[123] He and Avishai Margalit further explain that:

> By wearing a uniform, you take on yourself a risk that is borne only by those who have been trained to injure others (and to protect themselves). You should not shift this risk onto those who haven't been trained, who lack the capacity to injure; whether they are brothers or others. The moral justification for this requirement lies in the idea that violence is evil, and that we should limit the scope of violence as much as is realistically possible. As a soldier, you are asked to take an extra risk for the sake of limiting the scope of the war. Combatants are the Davids and Goliaths of their communities. You are our David.[124]

In essence, they are saying that soldiers must be willing to risk their own lives to save those of civilians, if necessary. As Plaw puts it, if at any given time there is a tradeoff to be made between combatants and noncombatants, any increased risk must be borne by the combatants.[125] In his exchange with Plaw, Asa Kasher objects to placing the burden of risk on combatants.[126] We are asked to imagine a scenario where military and political leaders, in planning a particular operation, are confronted with two possible means to achieve the same legitimate objective: one that endangers no civilians but puts many soldiers at risk; and one that endangers one civilian but only puts one soldier at risk.[127] If we take the positive commitment to save civilians' lives very seriously, he says it would seem to require the planners to choose the first option, because, while it risks the lives of many soldiers, it completely eliminates risk to civilians.[128] However, Plaw goes on to say that to require combatants to assume such risk in these situations under the 'due care' requirement significantly and unjustly devalues the soldiers' lives in relation to the lives of the enemy's civilian noncombatants. There is certainly some truth to this objection because, while soldiers are trained to kill and usually pose some risk to innocent people, to say that a citizen-soldier's life is less worthy than that of a civilian who has chosen not to fight in defence of the state and the common life seems highly problematic. If, for instance, a state has the option of using unmanned systems and this places the enemy noncombatants at some small risk, while sparing the lives of many soldiers, it would be difficult to justify the impermissibility of their use. However, as repeatedly noted throughout this chapter, risk to civilians and others on the ground will, in practice, be heightened in comparison to conventional combat. In this sense, the 'due care' problem, which is central to the matter of risk transference and morality of unmanned warfare, is exacerbated or complicated by the West's growing reliance on unmanned systems.

While acknowledging that there is an element of truth to Kasher's claim and that some might maintain that we must take the commitment to save civilian lives to be totally absolute, even Walzer suggests that there is a limit to the risk that states, military forces and their personnel are required to assume.[129] For him, we must simply accept that these deaths are collateral damage, cannot be appropriately eliminated and thus the states, their military forces and personnel are not obliged to go to great lengths to eliminate them. We cannot have an absolute prohibition against risking civilian lives because war necessarily places humans in danger. That is just one part of war's 'hellishness', as Walzer describes it.[130] Walzer therefore wants soldiers to assume some of the risk, but not all of it. Exactly how far they must go to protect civilians is hard to say. The matter, then, is really one of determining the level or degree of risk that can be shifted. For our part, it is clear that shifting too much of war's risk onto civilians is morally problematic, since military planners are likely to do this whenever they have the opportunity to do so. This, it will be argued in the next chapter, may also mean that they will be tempted to carry out more operations and will allow more civilians to be harmed. Walzer goes on to ask whether it might be ethically permissible to impose a one-in-ten chance of death on a civilian, while a three-in-ten chance is unjustifiable?[131] Tim Clement tries to determine whether there is a figure at which we can draw a line. He uses casualty figures from past conflicts to try to determine a limit of tolerability for risk to one's own forces during operations, but inference from what was once considered tolerable to what is now considered tolerable is unlikely to be the best strategy given the changing character of warfare.[132] The degree of risk must surely have much to do with the nature of the target, the threat they pose and, more so, the available technology. For this reason, it is quite difficult to argue that when it comes to unmanned systems, military forces hold anything more than a general obligation to reduce risk as far as is reasonably practicable.

Again, the expression 'as far as is reasonably practicable' is open to considerable interpretation. Walzer provides two well-known examples that help narrow down the range of acceptable risk. The first he draws from the memoirs of a World War I veteran, Frank Richards.[133] Richards was tasked with clearing cellars and dugouts during the battle for France. Prior to investigating the dark spaces, he would throw hand grenades in them to kill any hiding Germans. However, Richards was very worried about the presence of civilians. He adopted the practice of shouting down into the cellars in case there were any civilians taking cover in then. Otherwise, Richards thought he would be committing what he called 'innocent murder'. Of course, if there were German soldiers hiding out in the cellar, they would clearly come out shooting at Richards and his fellow soldiers, meaning that he (and his comrades) was taking on additional risk in shouting his warnings. The other example given by Walzer is the aerial bombing of occupied France during the Second World War. The Free French Air Force conducted bombing raids against military targets, but they knew that their bombs killed Frenchmen who were coerced into supporting the German war effort as well as others who simply lived in the vicinity. The pilots resolved the dilemma, not by stopping the raids, but by taking on greater risk to themselves. They flew at a much lower and riskier altitude, but this allowed for greater precision.[134] Walzer says the tolerable limit of risk is fixed, then, 'roughly at the point that any further risk-taking would almost certainly doom the military venture or make it so costly [in terms of soldiers' lives] that it could not be repeated'.[135] This idea, that warfighters must accept risk in order to protect civilians up to the point where they can no longer win battles fought in support of wars that must be won, makes a good deal of sense.

In many instances though, unmanned warfare does not seem to fit this Walzerian risk profile, because, in this specific type of warfare, the imperative of self-preservation tends to take precedence over strict adherence to the rules of war. Rather than making multiple transfers of risk, it would seem that the technologically advanced West should actually assume more risk and seek to develop technologies that remove eliminable risks rather than merely shifting them around. That is, they should put greater effort into ensuring that they have unmanned systems technologies and others that are safer or more robust, and that systems designed to keep people out of harm's way actually do so through the entire service of life a machine. This is only because the disparity in risk between the West and all others is so great, unlike in the earlier scenario. In a sense, the West, with its technological advantage, can afford to take greater time to guard against unnecessary harm. However, as we have already seen, there are few truly 'eliminable' risks in warfare, even with unmanned systems, so the West (as a whole) in assuming more risk will often involve soldiers assuming some extra risk to their lives for the sake of others. It is also worth noting here that there is an important difference in soldiers having to assume risk because of poor planning, a lack of foresight and the range of technical operations described earlier in this chapter, and soldiers assuming risk because it is righteous.

One might argue that, no matter how small or big the extra burden of risk may be, the requirement to take on any additional burden in utilising new weapons makes little sense and should therefore be dismissed; and that it is unrealistic to expect soldiers, who may be fighting for a good cause, to give up their right not to be harmed. However, it must be pointed out that this sort of argument need not rely on forfeiture theory. That is, it is not necessary to say that soldiers must give up their rights or anything of that nature. Many professionals have some sort of ethical obligation to risk their own lives to save those of others and we do not say that they have forfeited their rights.

As John Sullins points out, fire-fighters, police, ship's personnel and so on, all do this on occasion and there appears to be no reason why military personnel should be excused from this requirement.[136] Furthermore, if we were to remove this dictum from just war theory and permit states and their military forces to allow force preservation concerns to override any others through the endless transferring of risk, we would have to allow for the use of civilians and civilian buildings as shields. It is a slippery slope toward everything and everyone becoming legitimate targets of military force, and this will only lead to greater levels of violence and destruction, which is counter to the long-standing aims of just war theory and most unmanned warfare programmes.

Conclusion

It was suggested at the beginning of this chapter that while unmanned warfare, as a relatively new and primarily Western way of war, appears to improve on or transcend the moral deficiencies of earlier forms of warfare characterised by indiscriminate bombing campaigns and large numbers of combatant and noncombatant deaths, it contradictorily introduces new problems, new tensions, new paradoxes and new countervailing tendencies. These are numerous, as this chapter has indicated. To begin with, unmanned systems, which are supposed to keep people out of harm's way, involve many more people than is typically assumed, many in dangerous forward-deployed environments. These systems also encourage complacency and dependency in a sensor-saturated information society. This is problematic because it can result in states initiating or becoming entangled in wars to which they are not fully committed as well as in states slowly becoming unable to mount a robust military defence because of the displacement of military personnel, which is a key feature of unmanned warfare. Another consequence of these new weapons is that because morality and prudence call for weapons that are 'safe', it becomes much more likely that the military-industrial complex will demand 'test-bed' wars from the states that receive their financial support. Since unmanned platforms do not involve a human crossing onto foreign soil and invading another country's sovereignty, war may also be waged more frequently. Perhaps the biggest issue, however, is that the perceived precision that accompanies some unmanned systems technologies brings warfare into urban environments, which are military and morally challenging. Part of the challenge comes from the difficulty to distinguish between legitimate and illegitimate targets due to technical limitations and the intermingling of combatants, noncombatants and private military contractors. These problems are best understood in terms of a process of risk mitigation or management, as each involves a risk to human life which is been transferred to others, most problematically, noncombatant civilians. The morality of risk transfer in war is questionable at best and, in Chapter 6, we continue to look at the ethics and efficacy of unmanned warfare, venturing more explicitly into the psychosocial and political implications of unmanned warfare, exploring some further problems and a possible solution. However, the underlying argument is essentially the same: while unmanned systems have some obvious benefits, there are numerous technological and operational pitfalls for the unwary and these limit the moral obligation to use them and point to the need for further regulation.

Notes

1 Note that in the sense used here, this is a positive value as machines are said to accumulate the risk in lieu of human beings.

2 Gert-Jan Lokhorst and Jeroen van den Hoven, 'Responsibility for Military Robots', in *Robot Ethics: The Ethical and Social Implications of Robotics*, ed. Patrick Lin, Keith Abney, and George Bekey (Cambridge: MIT Press, 2012), 148 [emphasis added].

3 Keryl Cosenzo, 'Automation Strategies for Facilitating Human Interaction with Military Unmanned Vehicles', in *Human-Robot Interactions in Future Military Operations*, ed. Michael Barnes and Florian Jentsch (Farnham: Ashgate, 2011), 103 [emphasis added].

4 Benjamin Buley, *The New American Way of War: Military Culture and the Political Utility of Force* (New York: Routledge, 2008), 110 [emphasis added].

5 Robert Sparrow, 'Can Machines Be People: Reflections on the Turing Triage Test', in *Robot Ethics: The Ethical and Social Implications of Robotics*, ed. Patrick Lin, Keith Abney, and George Bekey (Cambridge: MIT Press, 2012), 305 [emphasis added].

6 Robert Sparrow, 'Building a Better Warbot: Ethical Issues in the Design of Unmanned Systems for Military Applications', *Science and Engineering Ethics* 15, no. 2 (2009): 173.

7 As with any machine, there is a risk that a particular component might fail and cause bodily harm to bystanders. With an unmanned aerial vehicle, for instance, a propeller may fail and wreak havoc on the ground crew. This is not the primary concern here.

8 John P. Sullins, 'Robowarfare: Can Robots Be More Ethical Than Humans on the Battlefield?', *Ethics and Information Technology* 12, no. 3 (2010): 264.

9 In the United States, these can be officers that have typically received training in just war theory and the law of armed conflict or members of the lower ranks (who often have not).

10 Micah Zenko, '10 Things You Didn't Know About Drones', *Foreign Policy* March/April(2012), http://www.foreignpolicy.com/articles/2012/02/27/10_things_you_didnt_know_about_drones?page=full.

11 Matt J. Martin and Charles W. Sasser, *Predator: The Remote-Control Air War over Iraq and Afghanistan: A Pilot's Story* (Minneapolis: Zenith Press, 2010), 30; 192.

12 Deployed outside of a safe-zone.

13 Much of Colonel Martin's book recounts his experience of 'going to war' in Iraq to train Italians, who had just acquired unmanned aerial vehicles and needed combat training prior to participating in allied operations in Iraq. See Martin and Sasser, *Predator: The Remote-Control Air War over Iraq and Afghanistan: A Pilot's Story*, 147–70.

14 Sparrow, 'Building a Better Warbot: Ethical Issues in the Design of Unmanned Systems for Military Applications', 173.

15 Other machines could conceivably rescue unmanned systems, but this capability has not yet been developed or used.

16 Martin and Sasser, *Predator: The Remote-Control Air War over Iraq and Afghanistan: A Pilot's Story*, 139.

17 Benjamin, *Drone Warfare: Killing by Remote Control*, 53.

18 Fred J. Pushies, *Night Stalkers: 160th Special Operations Aviation Regiment (Airborne)* (Minnesota: Zenith, 2005), 54.

19 In a footnote acknowledging the input of John Canning, Sparrow notes that building proper 'anti-tamper' measures into the system could offset some of the risks discussed here. However, one should further note that to completely eliminate the need to rescue downed systems, we would need a measure whereby the entire system would be blown up or otherwise destroyed in order to prevent reverse engineering. This is primarily a matter of designing these systems in a way that reflects practical matters of moral import. Sparrow, 'Building a Better Warbot: Ethical Issues in the Design of Unmanned Systems for Military Applications', fn 7.

20 Singer, *Wired for War: The Robotics Revolution and Conflict in the 21st Century*, 19–21.

21 Peter W. Singer, 'The Future of War', in *Ethical and Legal Aspects of Unmanned Systems: Interviews*, ed. Gerhard Dabringer (Vienna: Institute for Religion and Peace, 2010), 82.

22 Singer, *Wired for War: The Robotics Revolution and Conflict in the 21st Century*, 337.

23 Kate Darling, 'Extending Legal Rights to Social Robots', in *We Robot* (Coral Gables: University of Miami School of Law, 2012), 2.

24 See Lori Thompson and Douglas J. Gillan, 'Social Factors in Human-Robot Interaction', in *Human-Robot Interactions in Future Military Operations*, ed. Michael Barnes and Florian Jentsch (Farnham: Ashgate, 2011).

25 Ian Roderick, 'Considering the Fetish Value of Eod Robots: How Robots Save Lives and Sell War', *International Journal of Cultural Studies* 13, no. 3 (2010): 245.

26 The American Humvee and Russian AK-47 assault rifle are good examples.

27 D. Axe, *War Bots: How U.S. Military Robots Are Transforming War in Iraq, Afghanistan, and the Future* (Ann Arbor: Nimble Books, 2008), 1.

28 Noah Shactman, 'Robot Sick Bay', *Popular Mechanics* 183, no. 7 (2006): 18–19.

29 Dunlap, 'Technology and the 21st Century Battlefield: Recomplicating Moral Life for the Statesman and the Soldier', 13.

30 If we consider the earlier discussion concerning unmanned systems operators going to great lengths to rescue their systems, it easy to infer that the level of trust in these systems is quite high.

31 Sparrow, 'Building a Better Warbot: Ethical Issues in the Design of Unmanned Systems for Military Applications', 173.

32 Les Hauck, 'Have We Become … Complacent?', *Combat Edge* 11, no. 11 (2003): 10.

33 Edward Story, 'Complacency = Mishap', *Combat Edge* 15, no. 4 (2006): 11.

34 Michael Barnes and Arthur W. Evans, 'Soldier-Robot Teams in Future Battlefields: An Overview', in *Human-Robot Interactions in Future Military Operations*, ed. Michael Barnes and Florian Jentsch (Farnham: Ashgate, 2011), 18; Story, 'Complacency = Mishap'.

35 L. Booher and J. Appezzato, 'What Could Happen? No Room for Complacency in OIF', *Combat Edge* 13, no. 10 (2005): 6.

36 S. Miller, 'Technology and War', *Bulletin of the Atomic Scientists* 41, no. 11 (1985): 47.

37 Roy van den Berg, 'The 21st Century Battlespace: The Danger of Technological Ethnocentrism', *Canadian Military Journal* 10, no. 4 (2010): 14.

38 Ibid.

39 Stephen Biddle, *Afghanistan and the Future of Warfare: Implications for Army and Defense Policy* (Carlisle: Strategic Studies Institute, 2002), 32.

40 Note that this is a different (though related) argument to that explored in much greater detail in the next chapter, concerning the lowering of political thresholds to war.

41 Sparrow, 'Building a Better Warbot: Ethical Issues in the Design of Unmanned Systems for Military Applications', 173.

42 Ibid.

43 Kenneth P. Werrell, 'Did USAF Technology Fail in Vietnam?', *Airpower Journal* 12, no. 1 (1998): 87.

44 This is evidenced by the fact that they later had to develop the toxic 'Agent Orange' to destroy the foliage so that combatants could be spotted through the jungle canopy and air combat operations could be effective.

45 Singer, *Wired for War: The Robotics Revolution and Conflict in the 21st Century*, 316.

46 Lin, Bekey and Abney, 'Autonomous Military Robotics: Risk, Ethics, and Design', 83.

47 Krishnan, *Killer Robots: Legality and Ethicality of Autonomous Weapons*, 136.

48 Wayne Chappelle, Kent McDonald and Katharine McMillan, 'Important and Critical Psychological Attributes of USAF MQ-1 Predator and MQ-9 Reaper Pilots According to Subject Matter Experts', ed. Department of the Airforce (Ohio: School of Aerospace Medicine, 2011), 5–7.

49 Ibid.

50 Lin, Bekey and Abney, 'Autonomous Military Robotics: Risk, Ethics, and Design', 83.

51 Ibid.

52 Chappelle, McDonald and McMillan, 'Important and Critical Psychological Attributes of USAF MQ-1 Predator and MQ-9 Reaper Pilots According to Subject Matter Experts', 6.

53 Ronald Arkin, 'On the Ethical Quandries of a Practicing Roboticist: A First-Hand Look', in *Current Issues in Computing and Philosophy*, ed. A. Briggle, K. Waelbers and P. Brey (Amsterdam: IOS Press, 2008), 47–48.

54 Lin, Abney and Bekey, 'Robot Ethics: Mapping the Issues for a Mechanized World', 947.

55 For a general discussion of the prospects of technologically displayed workers and some comparisons with sweatshop ethics, see Rick Bookstaber, 'Will the Unemployed Really Find Jobs Making Robots?', http://www.businessinsider.com/will-the-unemployed-really-find-jobs-making-robots-2012-8.

56 J. M. Sullivan, 'Defense Acquisitions: DOD Could Achieve Greater Commonality and Efficiencies among Its Unmanned Aircraft Systems', ed. Subcommittee on National Security and Foreign Affairs and Committee on Oversight and Government Reform (Washington, DC: Government Accountability Office, 2010), 6.

57 Patrick Lin, 'Introduction to Robot Ethics', in *Robot Ethics: The Ethical and Social Implications of Robotics*, ed. Patrick Lin, Keith Abney and George Bekey (Massachusetts: MIT Press, 2012), 7.

58 Noah Shactman, 'Robot Canon Kills 9, Wounds 14', http://www.wired.com/dangerroom/2007/10/robot-cannon-ki/.

59 Ibid.

60 Lin, Bekey and Abney, 'Autonomous Military Robotics: Risk, Ethics, and Design', 77.

61 Colin S. Gray, *Another Bloody Century: Future Warfare* (London: Weidenfeld & Nicolson, 2005), 231.

62 If military robotics software is made open-source, as with much civilian software, the systems could be potentially be made somewhat safer. Of course, standards would have to implemented (and these could be the source of others problems) and care would have to be taken to ensure that potential enemies do not enter exploitative code.

63 Lin, Bekey and Abney, 'Autonomous Military Robotics: Risk, Ethics, and Design', 78.

64 Vladimir Slipchenko and Aleksandr Khokhlov, 'Shock and Awe: Russian Expert Predicts 500,000 Iraqi Dead in War Designed to Test Weapons', http://globalresearch.ca/articles/SLI303A.html.

65 Ibid.

66 Dwight D. Eisenhower, *Dwight D. Eisenhower: 1960–61: Containing the Public Messages, Speeches, and Statements of the President, January 1, 1960, to January 20, 1961* (Washington, DC: United States Government Printing Office, 1961), 1035–40.

67 Ibid.

68 Benjamin, *Drone Warfare: Killing by Remote Control*, 32.

69 On the latter, see Ray Moynihan, 'Selling Sickness: The Pharmaceutical Industry and Disease Mongering', *British Medical Journal* 324, no. 7242 (2002).

70 Benjamin, *Drone Warfare: Killing by Remote Control*, 32.

71 Ibid., 31–36.

72 Ibid., 44.

73 There is also a congressional 'unmanned systems caucus' that serves as an advocate for drones. See Jill Replogie, 'The Drone Makers and Their Friends in Washington', http://www.fronterasdesk.org/news/2012/jul/05/drone-makers-friends-washington/#.UBfn8FEWH0c. Congressional Unmanned Systems Caucus, http://unmannedsystemscaucus.mckeon.house.gov/about/purpose-mission-goals.shtml.

74 Slipchenko and Khokhlov, 'Shock and Awe: Russian Expert Predicts 500,000 Iraqi Dead in War Designed to Test Weapons'.

75 Ibid., 153–4; Roy Allison, 'Russia, Regional Conflict, and the Use of Military Power', in *The Russian Military: Power and Policy*, ed. Steven E. Miller and Dmitri Trenin (Massachusetts: MIT Press, 2004).

76 Jan Egeland, *A Billion Lives: An Eyewitness Report from the Frontlines of Humanity* (New York: Simon & Schuster, 2008), 18.

77 Cordesman, *The Iraq War: Strategy, Tactics, and Military Lessons*, 297.

78 This should not be confused with Rawls's ethical principle of the same name.

79 This argument for the testing of new military technology playing a role in *jus ad bellum* is also made by Coppieters and Kashnikov, 'Right Intentions', 89.

80 Zardari 2009, cited in House of Commons Foreign Affairs Committee, 'Global Security: Afghanistan and Pakistan, Eigth Report of Session 2008–09, Report, Together with Formal Minutes, Oral and Written Evidence' (London: Parliament: House of Commons: Foreign Affairs Committee, 2009), 75.

81 United Kingdom Ministry of Defence, 'Joint Doctrine Note 2/11: The UK Approach to Unmanned Aircraft Systems', Section 5–9.

82 Gregory F. Treverton, *Reshaping National Intelligence for an Age of Information* (Cambridge: Cambridge University Press, 2003), 85.

83 Robert Sparrow, 'Robotic Weapons and the Future of War', in *New Wars and New Soldiers*, ed. Jessica Wolfendale and Paolo Tripodi (Farnham: Ashgate, 2011).

84 Ibid.

85 Sarah Kreps and John Kaag, 'The Use of Unmanned Aerial Vehicles in Contemporary Conflict: A Legal and Ethical Analysis', *Polity* 44, no. 2 (2012): 270.

86 Buster C. Glosson, 'Impact of Precision Weapons on Air Combat Operations', *Airpower Journal* 7, no. 2 (1993): 4.

87 Ibid.

88 Phillip S. Meilinger, '10 Propositions Regarding Air Power' (Washington, DC: Air Force History and Museums Program, 1995).

89 Ibid., 41.

90 Michael Ignatieff, *Virtual War: Kosovo and Beyond* (New York: Metropolitan Books, 2000), 148–9.

91 Ibid., 149.

92 British Broadcasting Corporation, 'Belgrade Hit by Nato Blitz', http://news.bbc.co.uk/2/hi/europe/332328.stm.

93 Alice Hills, *Future War in Cities: Rethinking a Liberal Dilemma* (Portland: Frank Cass, 2004), 9.

94 Ibid.

95 Ralph Peters, *Fighting for the Future: Will American Triumph?* (Pennsylvania: Stackpole Books, 2001), 74.

96 Markus Wagner, 'The Dehumanization of International Humanitarian Law: Independently Operating Weapon System and Modern Armed Conflict', in *We Robot* (Coral Gables: University of Miami School of Law, 2012), 24.

97 See discussions in: Baker, *Just Warriors Inc.: The Ethics of Privatised Force*; Armin Krishnan, *War as Business: Technological Change and Military Service Contracting* (Farnham: Ashgate, 2008).

98 Lindsey Cameron, 'Private Military Companies: Their Status under International Law and Its Impact on Their Regulation', *International Review of the Red Cross* 88, no. 863 (2006): 576.

99 Ibid.

100 Rebecca J. Johnson, 'The Wizard of Oz Goes to War: Unmanned Systems in Counterinsurgency', in *Killing by Remote Control: The Ethics of an Unmanned Military*, ed. Bradley Jay Strawser (New York: Oxford, 2013), 163.

101 Ibid., 166.

102 Ibid.

103 Michael O'Hanlon, *Technological Change and the Future of Warfare* (Washington, DC: Brookings Institution Press, 2000).

104 Tyler Wall and Torin Monahan, 'Surveillance and Violence from Afar: The Politics of Drones and Liminal Security-Scapes', *Theoretical Criminology* 15, no. 3 (2011): 243.

105 Kreps and Kaag, 'The Use of Unmanned Aerial Vehicles in Contemporary Conflict: A Legal and Ethical Analysis', 261.

106 Johnson, 'The Wizard of Oz Goes to War: Unmanned Systems in Counterinsurgency', 17.

107 Wall and Monahan, 'Surveillance and Violence from Afar: The Politics of Drones and Liminal Security-Scapes', 246.

108 Sparrow, 'Robotic Weapons and the Future of War', 119.

109 Christian Enemark, 'Drones over Pakistan: Secrecy, Ethics, and Counterinsurgency', *Asian Security* 7, no. 3 (2011): 229.

110 Peter Bergen and Katherine Teidmann, 'Washington's Phantom War: The Effects of the U.S. Drone Program in Pakistan', *Foreign Affairs* 90, no. 4 (2011): 13.

111 Peter Bergen and Katherine Teidmann, 'The Year of the Drone: An Analysis of U.S. Drone Strikes in Pakistan 2004–2010', *Foreign Policy* (2010), http://www.foreignpolicy.com/articles/2010/04/26/the_year_of_the_drone.

112 Aleem Maqbool, 'Mapping US Drone and Islamic Militant Attacks in Pakistan', *BBC News South Asia* (2010), http://www.bbc.co.uk/news/world-south-asia-10728844.

113 Avery Plaw, Matthew Fricker and Brian Glyn Williams, 'Practice Makes Perfect?: The Changing Civilian Toll of CIA Drone Strikes in Pakistan', *Perspectives on Terrorism* 5, no. 5–6 (2011): 58–61; for further figures, see Avery Plaw, 'Counting the Dead: The Proportionality of Predation in Pakistan', in *Killing by Remote Control: The Ethics of an Unmanned Military*, ed. Bradley Jay Strawser (New York: Oxford, 2013).

114 Ulrich Beck, *World Risk Society* (Cambridge: Polity, 1999). 51.

115 Coker, *Ethics and War in the 21st Century*, 137.

116 Ibid.

117 Ibid., 138.

118 Martin Shaw, *The New Western Way of War: Risk-Transfer War and Its Crisis in Iraq* (Cambridge: Polity Press, 2005).

119 Ibid., 94.

120 Ibid.

121 Krishnan, *Killer Robots: Legality and Ethicality of Autonomous Weapons*, 122.

122 In a similar light, Cook argues that a member of the military forces sent overseas on a humanitarian mission could argue that this 'isn't what I signed up for' and that they would therefore be entitled to use the best force protection possible before they can return home. See Cook, *The Moral Warrior: Ethics and Service in the U.S. Military*, 125.

123 Walzer, *Just and Unjust Wars: A Moral Argument with Historical Illustrations*: 156.

124 Avishai Margalit and Michael Walzer, 'Israel: Civilians & Combatants', *The New York Review of Books* 56, no. 8 (2009), http://www.nybooks.com/articles/archives/2009/may/14/israel-civilians-combatants/.

125 Asa Kasher and Avery Plaw, 'Distinguishing Drones: An Exchange', in *Killing by Remote Control: The Ethics of an Unmanned Military*, ed. Bradley Jay Strawser (New York: Oxford, 2013), 55–6.

126 Ibid., 56.

127 Ibid.

128 Ibid., 56–7.

129 Walzer, *Just and Unjust Wars: A Moral Argument with Historical Illustrations*, 156.

130 Ibid.

131 Ibid.

132 Tim Clement, 'The Morality and Economics of Safety in Defence Procurement', in *Safety-Critical Systems: Problems, Process and Practice*, ed. Chris Dale and Tom Anderson (Springer: Dordrecht, 2009).

133 Walzer, *Just and Unjust Wars: A Moral Argument with Historical Illustrations*, 152.

134 Ibid., 157.

135 Ibid.

136 Sullins, 'Aspects of Telerobotic Systems', 266.

Chapter 6
Unmanned Warfare: The Moral Costs of Changing Mindsets

In keeping with the risk-economy theme developed in the previous chapter and building on the technological and operational dimensions already discussed, this chapter will examine the efficacy of unmanned systems. In particular, it will focus on the mindset-altering dimensions of unmanned warfare and their impact on the principal war-making agents, namely: the public, unmanned systems operators, their commanders, higher-level military and political decision makers. The reason for this focus is explained by the fact that many of the unintended effects of this technology cannot be attributed to the machine, as per the previous chapter, but to human psychology. The first section will argue that the risk transference, which is characteristic of Western unmanned warfare, leads people to think that employing unmanned systems technologies allows us to wage less devastating wars and, as a consequence, has the concerning effect of making it too easy to wage war. In, other words, unmanned warfare challenges the principle of right authority and increases the propensity to use force. In the second section, we will look at some further problems associated with technologically mediated fighting and suggest that, through a process of moral disengagement and desensitisation, the barriers to immoral conduct *in* war may also be reduced. Having considered the impact on the long distance warrior's capacity or willingness to adhere to *jus in bello* norms, the third section will discuss the impact on the personal wellbeing of the operators themselves. Here, the arguments will centre on the idea that the impact of being simultaneously present in contrasting environments, i.e. where the war is fought and where one operates the unmanned system, may lead to serious transgressions of some just war principles. In the fourth and final section, we will concentrate on whether we can eliminate or relieve some of these technologically mediated, but distinctly human, moral problems by automating elements of the decision-making process. This chapter will conclude that, while greater automation certainly has the potential to alleviate some moral concerns generated by these systems, there is a strong case for keeping humans in the decision-making chain, even if it involves having to make a delicate moral trade-off between maintaining and/or improving warfighting capability and limiting harm to noncombatants.

Right Authority and the 'Threshold' Problem

George Lucas writes that one of the chief criticisms that can be levelled against unmanned systems, like many other recent technological innovations in war, is what he terms the 'threshold problem'.[1] As Lucas presents it, the threshold problem incorporates many of the arguments put forward by critics of these innovative military technologies (including Sparrow,[2] Singer[3] and Lin et al.[4]), but he narrows it down to two relatively distinct formulations. The first states that insofar as all such technological innovations tend to render war itself less destructive and costly in terms of the loss and injury to human life, property and the environment, they may also inadvertently lower the threshold for resorting to war and thus undermine the following *jus ad bellum* principle: war should only be resorted to as a method of alleviating or eliminating conflict as a last resort, i.e. when other nonviolent options have been exhausted or are too impractical.[5] This particular statement of the threshold problem raises some deep philosophical questions, which will be deferred until the next chapter. In this chapter, we will only be concerned with the second variant of the problem at the *ad bellum* level. This variant states that, insofar as unmanned systems technologies disguise the true costs of war, or otherwise appear to minimise these costs, and as a result alter the public mindset, they enable high-level war-makers (e.g. political and military leaders of liberal democratic governments) to engage their military forces in warfare without the full knowledge or procedural consent of those they govern. This surely leads to rejecting some interpretations of the principle of right authority.[6] Peter Asaro believes that this second variant – which he thinks makes it somehow easier for leaders to take an unwilling nation to war – is one of the strongest moral objections to the development and use of unmanned systems.[7] As will be explained, this author's inclination is much the same, not only because it creates problems for international society, but also because it violates both the basic social contract on which governments are formed and the military-state contract that governs the armed forces as their instruments. It will be argued here that employing unmanned systems and distancing warfighters from the conflict zone changes the public mindset, making it easier for our political and military decision makers to secure consent for what are potentially unjust wars. More specifically, it will be contended that there is a corrosive effect at work, not just on the actors and decision makers, but on the 'legitimators' whose consent is to be assumed, sought or provided.

While critics argue that just war theory should not be applied to war waged with modern technology, it was argued in Chapter 4 that the theory is very much relevant and limits the kinds of war that armed forces can wage in the name of democratic states. Ignatieff writes that the power to give or withhold consent for war – that is, to authorise war – is an essential element of the freedom of citizens, a freedom from which they cannot easily be alienated.[8]

War and the provision for defence against internal and external threats are also core functions of any legitimate state. As argued in Chapter 3, under the social contract, members of these states cede to the collective national institutions their individual right to use force, a right which they would otherwise possess in the state of nature. This means that there is a limited number of cases in which providing for one's own personal defence is the most practicable and defensible option. In any other case, the contracting parties agree to refrain from waging private wars and accept that the right or responsibility to resort to war and utilise mass violence is one reserved exclusively for the state, which putatively represents the collective. To protect the rights of individuals, the principle of right authority exists to regulate and legitimate any necessary use of force as well as to ensure that the power to authorise the deployment of weapons (such as unmanned systems) rests only with certain authorities. The requirement holds that states are permitted to wage war only if, in addition to meeting all other *jus ad bellum* criteria, the conflict has been approved through all legitimate channels that have been established over time to meet the requirements of the social contract. This authorisation, which is needed by the state's military forces, is typically embedded in domestic legal codes and constitutions, international law and/or sought from the United Nations.

Ignatieff explains that in modern practice, these codes, constitutions and institutions usually codify that, when countries face a threat, their leaders have the power to make a declaration of war only after having sought approval for that declaration (or subsequent funding) from their citizens' elected representatives.[9] In the case of the US, the President traditionally declares war and then seeks formal authorisation from those elected as representatives in the Congress. However, as Ignatieff notes, for almost half a century these channels have often been bypassed by Western states that have sent their soldiers into armed conflict without having sought proper authority beforehand. He cites Bosnia, Haiti, Iraq (on both counts), Kosovo, Panama and Somalia as examples.[10, 11] This is not to say that there was no consultation, but rather that the consultation was superficial and conducted under the guise of transparency. Singer finds this sort of subversion of the right authority requirement morally unacceptable and says that, for a given conflict waged by a democratic state to be considered justified, it must both have been approved through the legal channels established under the social contract, and involve the input and overall support of the general public. Otherwise, he says, the following aspects are perverted: the link between the public and the state's defence policy, the conflict resolution options that are available for pursuit, and the whole idea of the democratic process and citizenship.[12] In his view, the power to give or withhold consent is retained under the social contract. Such power can and ought to be more robustly exercised when it comes to authorising war. It appears that Singer wants to develop a stronger standard of right authority, i.e. one that involves a broader deliberative threshold that must be reached in order for the authority to wage war to be conferred.[13] For instance, a certain portion of the public writ large might have to support the action to ensure that it more accurately reflects the moral character of the community. That is, given the moral gravity of the potential outcome of war, it may not be enough for citizens to defer judgment to their elected officials.[14]

A more demanding requirement for right authority may be an improvement compared to the existing moral standard and could bring about positive change and perhaps fewer less devastating wars. After all, wars that receive strong public support conceivably stand a greater chance of ending faster and more successfully than those that do not. This is because governments can divert more troops as well as more funds, which will be used to provide the troops deployed with more and/or better equipment and weapons. Also, public support undoubtedly has an impact on the morale of deployed forces. Vietnam and Iraq are both good examples of conflicts that did not receive such support and resulted in what many would consider defeat and humiliation as a consequence. However, the relevant concern here is not with the right authority requirement itself or the earlier mentioned tendency to directly bypass democratic channels, but rather with the thought that unmanned systems further compound and complicate adherence to the requirement by altering the psychological mindset of the public. This is namely by making it easier for military and political leaders to attempt to garner public support for their chosen conflicts by engaging in a range of deceptive practices, including what Ignatieff calls 'linguistic subterfuge'.[15] The word 'war', he says, is the last word that a key political decision maker wants to use.[16] Anthony Cordesman, another military expert, has noted as well that 'one of the lessons of modern war is that war can no longer be called war'.[17] Since citizens are the ones who have to fight wars under the social contract as well as bear the impact on the economy and their lives, they tend to be more sceptical of war than their leaders. For most, the word 'war' is associated with the prolonged and deadly conflicts of earlier generations, so instead, military and political leaders use words like 'hostilities', 'precision strike', 'coercive diplomacy' and other synonyms for war. This seems to disguise the moral reality of the planned or undertaken action/s, further altering the public mindset. In truth, even unmanned warfare involves human forces taking fire or being put at risk and subjecting others to the same. War makers often also utilise polls and focus groups in order to gauge public support for their desired actions and craft a convincing public appeal for action.[18]

The costs of war, whether measured in terms of human or political capital, also tend to mount much more quickly than is usually anticipated. Nevertheless, Western leaders continue to sell warfare to the public as risk-reduced (or even risk-free) and, as argued in the previous chapter, utilise multiple transfers of risk to manipulate the public into granting them authority to wage war, despite the fact that these transfers often only shift or introduce new forms of death, disaster and failure.[19] One might have thought that with the media's power and omnipresence, the anticipated or ongoing costs of any particular war would have be realised quickly, thus allowing the public to make an informed decision. However, truth is often war's first casualty, even in the modern era. As Phillip Knightley reveals in his book on the subject of the war correspondent as a myth-maker, states and their military forces have become much more adept at managing public opinion via the media since Vietnam.[20] In fact, some theorists go so far as to argue that, in the modern age and with its new weapons, the media has 'indisputably become a weapon of war'.[21] While a military's ability to control the media is limited to some extent by ethical and legal considerations, military forces have long recognised the role of the media in shaping the public mindset. As a result, military forces may exercise their ability to limit access the battlefield during violent conflicts in order to exert control over what information is divulged and to whom it is released (limited by ease of transferring digital data).[22] For instance, experience gained in Iraq has taught the US military that graphic images can have a debilitating effect on the public's support for war such that they cannot carry out operations in the way they would like to.[23] As such, military forces have sought to distance the public from the horrors of war for many years by banning photographies of the coffins of deceased soldiers.[24] The apparent aim in attempting to control the media is to morally disengage the public from war's harsh realities in the hope that they will more readily support military action. In this sense, to engage in media and risk management is really to engage in yet another form of mindset-altering politico-moral subterfuge.

A possible objection to this line of argument, at least as it applies to unmanned systems, is that these weapons actually improve public understanding of the horrors of war. These systems typically have high-resolution optics and other equipment that allow them to accurately and clearly record events and conditions. Whether it is the military forces themselves for propaganda reasons, or the soldiers and other whistle-blowers who feel the need to share their experiences, footage taken from these systems has often found its way into the public arena. The effect of this drone footage is compounded by the fact that throughout the Iraq war, amateur videos of combat produced and posted by soldiers on video sharing websites such as *YouTube* and *MySpace* proved to be a ready source of humiliation for the US military.[25] In 2007, the Defense Department forbade military personnel from uploading similar material to the Internet, but the prohibition was reversed two years later, when it came to realise that, while a military force is able restrict the media's access to the battlefield, it does not have full-spectrum control over the actions of its own troops.[26] There is now a 'large number of video clips showing members of the US (and British) military engaged in extremely violent, anti-social activities'.[27] The argument could be put forward by the advocates of remote weaponry that the raw footage from unmanned systems, coupled with that from phones and other personal technologies, makes it significantly less likely that the public will underestimate the moral gravity of war or otherwise be swayed toward supporting or accepting a particular engagement.

However, it remains to be seen whether this is the case. Benjamin vehemently opposes such claims, arguing that while the footage may well expose the public to the horrors of war, it somehow makes it look 'fun'.[28] She says that the video clips from drones, which are often edited to various sound tracks and captions, turn war into a form of entertainment that has been termed 'war porn'.[29] Many show Americans killing 'faceless' soldiers, which only serve to reinforce racial stereotypes and encourage alienation. A sampling of the comments attached to these videos gives an indication of their effect on the public and their willingness to confer authority to wage warfare even after having seen them:

> Fuk [*sic*] the Iraqis and look at the bigger picture. We're in a battle for civilization here.[30]

> Let the bodies hit the floor! Let the bodies hit the floor! Let the bodies hit the floor. I want Muslim blood!!![31]

> I love the smell of burnt Muslim in the morning. They better pre-order some of them virgins [*sic*] for these murderous terrorist jihadists! Allah Kaboom! I'm loving this shit – blow them mutha fucka's [*sic*] to pieces.[32]

A further indication of the popularity and influence of the videos is reflected by the fact that some of them have received as many as 10 million unique views.[33] If the thousands of other videos, their comments and number of views are indeed accurate indicators, it would seem that unmanned systems actually counter what David Whetham labels the 'CNN effect', whereby public support for war is expected to fall as civilian casualties eventually mount and friendly forces as well as noncombatants become casualties in increasing numbers.[34] Of course, it may be that a large segment of the public is much more conflicted. David Betz and Sangho Lee argue that what is likely to emerge is a cognitive dissonance in the public mind.[35] This, they say, is due to being morally and physically disengaged from the battlefield while simultaneously being able to view the harm and suffering caused in a level of detail that was rarely available to the public in earlier conflicts.[36] On the one hand, the public is told that their armed forces are utilising unmanned systems to conduct relatively safe and peaceful operations. But, on the other hand, the very same public is presented with dreadful imagery of its own soldiers dead, wounded or, as previously mentioned, paraded before the international media. All of this arises amidst a barrage of scenes detailing the unfortunate fate of civilians caught somewhere in the crossfire. Regardless of whether unmanned weaponry results in some sort of moral disengagement or a form of public cognitive dissonance, the risk comes in generating a liberal democratic citizenry that is in some sense and to some degree psychologically divested of its moral right or ability to confer authority to wage war in accordance with just war theory. A disengaged or conflicted citizenry is unlikely to be able to balance the need to protect the community without unjustly threatening other communities.

At first sight, this general concern about the implementation of unmanned systems appears to be one of the strongest. Indeed, as Strawser writes, the worry that by utilising unmanned weaponry, we might alter the public mindset and make it too easy to gain the consent of the governed, thus making the waging of war more likely, is one that seems to hold much real-world relevance and plausibility.[37] However, he argues that the right authority threshold argument is not as strong as it at first seems, because it appears to obscure a logical implication following from it, i.e. that we should keep war thoroughly within the realm of the horrifying due to the deterrence effect that such horror provides. American General Robert E. Less affirmed a similar sentiment at the height of the one-sided Battle of Fredericksburg when, upon witnessing surges of Union soldiers charging toward what was certain death and destruction, he turned toward his second-in-charge and said that 'it is well that war is so terrible [or] we should grow too fond of it'.[38] Lin also interprets the argument as saying that we should not do anything that makes armed conflict more palatable, whether that be seeking to reduce casualties or introducing new weapons technology, which eliminates or transfers risk to others or, more generally, improving things like battlefield medicine or performing further research to improve operational outcomes.[39] Quite strikingly, he goes on to say that taken to the extreme, the argument seems to imply that we should actually take active efforts to raise barriers to war and make it as horrible and savagely violent as we possibly can.[40] Both he and Strawser claim that this argument might therefore be taken to suggest that we should block the development of any military technology that creates an asymmetry in warfare. This argument could also be employed to demand that military forces revert to warfare fought with primitive weapons as well as go without armour so as to ensure that both citizens and political and military decision makers become as reluctant as possible to make the initial decision to wage war in the hope of minimising incidences of unjust war and overall harm and destruction.[41]

This counter-argument seems ill conceived against unmanned systems and it lacks relevance when put into a real-world context. To see why, it will help to examine a supporting example provided by Strawser. He asks us to consider two possible worlds: Alpha and Beta.[42] In the Alpha world, a nation-state called Zandar has developed its technology and has the resources to manufacture bulletproof vests for its military personnel to wear in combat (which significantly reduce their risk of dying). In the Beta world, Zandar has developed the very same technology and resources. It is also involved in the same violent conflicts as Alpha-Zandar. However, Beta-Zandar reasons that, if it utilises bulletproof vests, it would, to some extent, reduce the cost of war in terms of human lives, thus making war more palatable and more tempting to wage. Beta-Zandar also reasons that, given the difficulty that can accompany determining the justice of war, some wars will be unjust. As such, it decides against using the bulletproof vests in order to make war more deadly and horrifying. This is done in the hope that it will lessen the likelihood that Zandar will engage in an unjust war in the future Beta world.[43] If Beta-Zandar could, indeed, be sure that, by foregoing the bullet-proof vests, it would be involved in fewer wars (particularly unjust wars), then there would be nothing counterintuitive about doing so, or in abolishing some other military technologies. Perhaps there should be no weapons at all, nor perhaps military forces. Perhaps if all other states would also relinquish their weapons, this would be the best way to achieve a better state of peace. It would, however, be competitively disadvantageous and morally reckless for a state (in our present world) to relinquish its military technology in the hope of making it more difficult to secure consent for war. Indeed, doing so would be to rely on what Strawser calls 'epistemically dubious calculations' about the future.[44] It is counterintuitive to take some grave risk now out of concern for some unlikely future event. Of course, this is not to say that states should absolve themselves of any concern or responsibility to accept risk in offsetting possible undesirable outcomes. Strawser appeals to extreme cases in an attempt to dismiss the problem, but the use of unmanned systems does not logically entail that we will be taken to the extremes of Zandar in the Beta world. In reality, we ought to balance important risks about future events with the moral consequences of making war more palatable. For instance, if it is clear that unmanned systems have the potential to undermine the right authority requirement, provisions need to be put in place to counter any potential consequences. It is, in this sense of avoiding more devastating war in the form of fewer deadly wars, or less frequent but more deadly wars, that proposing a more demanding right authority requirement seems the logical course of moral action.

Before considering how unmanned systems make it easier to conduct war once the decision to wage it has been made, let us consider one final argument against strengthening the right authority requirement. Julian Savulescu and Zack Beauchamp defend the view that lowering the threshold to engagement in war is not necessarily a morally bad thing, contrary to what is commonly assumed. They argue that in at least one case – that of humanitarian military intervention – altering the public mindset to make it easier to gain consent for military operations is in fact a positive feature of the use of unmanned weaponry.[45] They say that states often fail to engage in peacekeeping operations and humanitarian interventions, even when they are justified in doing so.[46] One clear reason for this is the public's aversion to casualties. There are varied reasons for this casualty aversion, some of which have already been discussed. In the United States, for instance, it could be a result of the Vietnam legacy or simply a reaction to the 'strategy of grinding attrition', which characterises many of the West's past wars.[47] Savulescu and Beauchamp say that this casualty aversion can be morally problematic for two reasons. One, it can result in certain risks being unjustly transferred to other parties. Two, states could be morally obligated to get involved in some conflict and may be blameworthy if they decide against launching some sort of intervention.[48] Therefore, they argue that, if intervention is morally permissible or obligatory, and if unmanned systems do make it easier for states to intervene in humanitarian cases and mitigate or resolve these two concerns, then it is surely a good thing that unmanned systems make it somewhat easier to overcome this potentially morally problematic mindset, driven by casualty aversion, and to secure the consent of the governed in order to intervene.[49] In other words, their argument is that a state's subversion of the right authority requirement might inadvertently yield some good.

However, this does not fix the underlying problem, which, according to their own argument, has more to do with the West's desire to unjustly prioritize its own citizens and soldiers over others, e.g. the citizens of Rwanda and Darfur. Furthermore, the power of the right authority requirement comes from its ability to prevent unjustified conflicts, but Savulescu and Beauchamp fail to explain how, with the weak interpretation of the principle they seem to favour, they intend to ensure that unmanned systems will only be used in justified interventions as 'robot guardians' (as they call them). While states will often fail to intervene when they ought to, they much more frequently intervene when they should not. The reality is that foreign intervention in the domestic affairs of states will, in some circumstances, only shed more blood. To use a current example, as countries like Syria descend into civil war and levels of human misery rise, pressure for yet another western military intervention in the Middle East continues to grow. But it is not at all clear that high-technology military intervention is the most suitable and morally sanctioned course of action for resolving civil wars, which tend to have the support of the local population. Right authority essentially gives people power of veto over the decisions of the state and its military institutions in an effort to ensure that an appropriate balance is reached between intervention and non-intervention in these unclear cases. For this reason, there is reason to oppose Savulescu and Beauchamp's view and believe that a strengthening of the right authority requirement or a much more careful interpretation of what it demands – both in terms of humanitarian military intervention and regular war – is called for in a military environment such as ours, i.e. in which unmanned platforms and other mindset-altering and potentially threshold-lowering technologies exist.

Lowering Barriers to Killing in War: Discrimination and Proportionality

Until this point, we have considered the various ways in which unmanned systems alter the public mindset and make it easier for military and political leaders of liberal democratic states to take an unwilling nation to war, undermining respect for the *jus ad bellum* principle of right authority. As already mentioned, this concern about the use of unmanned weaponry is perhaps one of the greatest that faces us, but while the high-level decision makers, who are central to this concern, are responsible for making the initial decision to engage in warfare, it is the individual soldier who defends his or her state and society who must be most unconditional in their exercising of moral restraint and in their adherence to just war theory. Ignatieff writes that it is the soldiers, more than any other war-making agential group, who have the most influence on the war's outcomes as well as the ability to introduce a moral component, because they actually conduct the war in question. In his words, 'the decisive restraint on inhuman practice on the battlefield lies within the warrior himself – in his conception of what is honourable or dishonourable for a man to do with weapons'.[50] Ironically, soldiers are the primary agents of physical violence, compassion and moral arbitration in war. As Darren Bowyer remarks, they deliver 'death and destruction one moment … [and deal] out succour to the wounded (of both sides) and assistance to the unwittingly involved civilian population, the next'.[51] The specific concern examined here is whether by removing soldiers from the battlefield and training them to fight via a technologically mediated proxy, we may, through a process of psycho-moral disengagement and desensitisation, lower their ability or willingness to exercise restraint and compassion in warfare as well as their adherence to the *jus in bello* principles of discrimination and proportionality. It will be argued that the employment of unmanned systems is linked to unethical decision-making and/or lowers barriers to killing, thus endangering the moral conduct of warfare.

To understand the relationship between drones, distance and the lowering of barriers to killing, we first need to understand how soldiers come to be able and willing to kill. Most, if not all, human beings are born with what can only be described as a primitive survival instinct that, without unchecked force, would lead to a certain degree of violence and savagery. But in most societies, people are raised and socialised in such a way that typically leads them to hold an aversion to harming other human beings. In a military context, this socialised reluctance to kill is evidenced by recounts and statistics from earlier wars. David Grossman, a self-proclaimed 'killogist' or military psychologist, uses the cases of two World War veterans.[52] The first confirms that many World War I infantrymen never fired their weapons and relied instead on artillery. The second relates that platoon sergeants in World War II had to move up and down the firing line kicking men to get them to fire and they felt they were doing good if they could 'get two or three men out of a squad to fire'.[53] While some have criticised his methodology, S. L. A. Marshall gives further supporting evidence when arguing from a mixture of personal experience and studies conducted on firing ratios. These reveal that 'on average not more than 15 per cent of the men had actually fired at the enemy'.[54] He attributes this startling inhibition to kill to an 'ingrained fear of aggression' that was based on society's teaching that killing is fundamentally wrong.[55] For Marshall, success in combat and the welfare of the state and its people, demands that action be taken to correct or overcome this problem.

In the years following the publication of the first edition of Marshall's book – that is, in the years following World War II – there is evidence that Marshall's call for corrective action were answered and that we were put on a worrying trajectory toward our unmanned systems dilemma. The evidence of very low firing rates has been replaced by evidence of very high and morally concerning firing rates. By the time of the Korean War, the American firing rate was said to be up by 55 per cent and, in Vietnam, it was reported to be up by 90 or 95 per cent.[56] Some express doubts about these firing rates too, as some troops in the rear of the formation were found with unspent ammunition, but those doubting were generally satisfied with the fact that among those who actually sighted the enemy, there appeared to have been extraordinarily high and consistent firing rates.[57] From a strictly military or operational perspective, this is a remarkable success story. In order to overcome the hesitancy to fire and kill that most people develop over time, Russell Glenn says that staff sergeants and platoon commanders watched their troops to ensure that they were actually engaging with the adversary. In Vietnam, they also listened for the steady roar of machine gun fire, an indication that their soldiers were unhesitatingly firing their weapons.[58] However, this corrective action seems unlikely to account for such a radical shift in the firing ratios. The real cause for the difference in the firing rates, it is argued here, has much more to do with the technology employed in later conflicts alongside changes in military training which, together, allowed and continues to allow, individuals to achieve a physical, emotional and/or moral distance from their enemies, thus enabling them to kill with more ease. It is these distances that need to be explored in more detail, since unmanned systems only make them grow as well as the disengagement and the accompanying desensitisation to a point that gives rise to unique problems affecting the operators' ability to wage discriminate and proportional warfare.

The link between physical and emotional distance, ease of aggression and waging warfare is in no way a new discovery. As Grossman writes, it has long been understood that there is a positive relationship between the following aspects: the empathetic, the spatial proximity of the victim, the difficulty to kill, the personal trauma caused by the kill and, more generally, the morally problematic ease of killing.[59] This relationship has been a cause for concern among anthropologists, philosophers, psychologists, theologians and, of course, soldiers themselves, who often struggle to understand their own actions. Jesse Glenn Gray, an American philosophy professor whose career was interrupted by a period of service as a World War II counter-intelligence officer, wrote that unless one is caught in some sort of overwhelming murderous ecstasy where rage takes over, killing and destroying is much easier when done at a distance and that with every foot of distance, there is a corresponding decrease in the accurate portrayal of reality.[60] He remarks that there is a point at which one's representation of the world begins to flag and another at which it fails altogether.[61] Glenn noticed this over 50 years ago. The concern is that current unmanned systems seem to increase the relevant distances to the point he referenced, at which a sense of reality is weakened. Even worse than that, the advancement in unmanned systems technologies may reach a point whereby the moral inhibitions of unmanned systems operators will be almost totally overcome. That is, the worry is that we have reached a morally significant junction in the history of weapons development. To properly understand how these weapons make it easier to kill and how they dampen a soldier's capacity to act ethically in war, we have to think about unmanned systems in the wider context of weapons that increase physical distance to the target and lower resistance to killing.

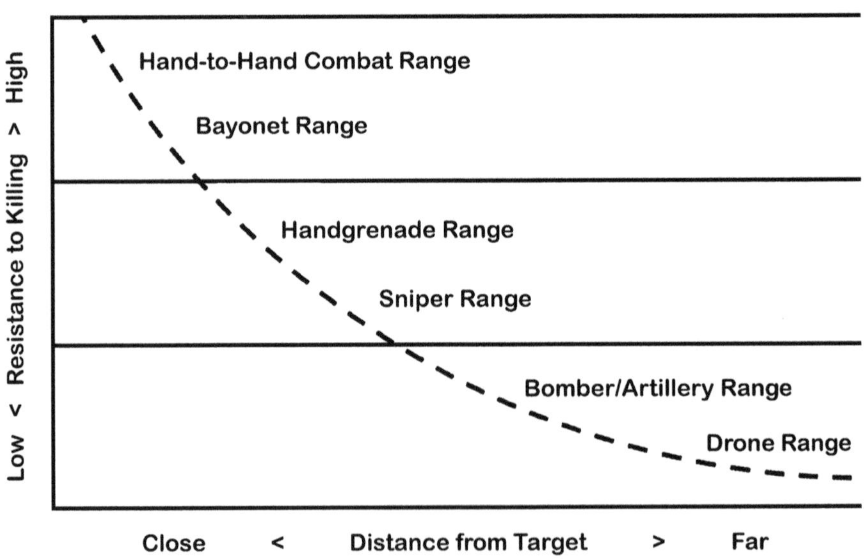

Figure 6.1 The relationship between killing and physical distance: From hand-to-hand combat to drone range

The figure above, adapted from Grossman,[62] depicts the phenomenon described above. It plots various weapons ranges and shows the hypothesised correlation between distance from target and the resistance to/ease of killing. The horizontal lines demarcate three roughly divided killing ranges (from top to bottom): close range, midrange and long range. Close range, for our sake, involves any easily-attributable kill at 'point-blank' range, whether with one's bare hands, an edged weapon or even a projectile weapon. According to Grossman, the key factor in close range killing is the undeniable responsibility one holds for the act. John Keegan and Richard Holmes cite the story of an Israeli paratrooper during the capture of Jerusalem in 1967: 'we looked up at each other for half a second and I knew that it was up to me, personally, to kill him'.[63] When a soldier kills at this range, more than any other, it is an intensely vivid and personal matter.[64] One can see the raw emotions on their enemy's face, hear their cries and smell the gunpowder. The Israeli paratrooper mentioned above goes on to say that having shot his enemy at close range, he could see the hate on his enemy's face and that 'there was so much blood … [he] vomited'.[65] Combat at close proximity is an interpersonal affair, so much so that it is incredibly difficult to deny the enemy's humanity. For this reason, Grossman says, the resistance to close-range killing is 'tremendous'.[66] At midrange – where you can still see and engage the enemy with hand grenades, sniper rifles and so on, but usually without being able to gauge the extent of the wounds inflicted – the experience of killing changes. At this range in the spectrum, the soldier can deny responsibility for the fatal shot or blow if there are others present and participating in the act of killing. One is still located on the battlefield and can hear the gunfire and feel the stress, but the distance between adversaries makes the act of killing both physically and psychologically easier, and thus potentially more morally troubling.[67]

At long range, where one must use some sort of mechanical or electrical assistance to view potential victims (i.e., binoculars, cameras or radar), there is evidence to suggest that killing is made even easier. Among those who are least reluctant to kill are pilots, artillery numbers and missile silo attendees. Gwynne Dyer writes that, while being observed by their fellows puts pressure on them to kill (as was the case with the gunners in Vietnam), their reduced reluctance has much more to do with the distance between them and their targets, which somehow acts as a moral buffer. She aptly notes that on the whole, 'gunners fire at grid references they cannot see; submarine crews fire torpedoes at "ships" (and not, somehow, at the people in the ships); and pilots launch their missiles at "targets"'.[68] Grossman also reports that in his extensive career researching and reading on the subject of killing in combat, he is not aware of a single instance in which an individual operating at such long range has refused to kill.[69] We have numerous examples of long distance killing made easy. Dyer reminds us that in the early 1940s, for instance, the British Royal Airforce 'firebombed' Hamburg.[70] Using early bomber aircraft, munitions blew in windows and doors over four square miles and resulted in a firestorm which left 70,000 people dead, mostly women, children and the elderly.[71] A further 80,000 died in the firebombing of Dresden, 225,000 in Tokyo and many millions more in bombing conflicts since.[72] If the bomber crews had to kill each of these people with a flamethrower or, as Whetham writes, slit each of their throats with a knife, the majority would be unable to do it.[73] The awfulness of killing people at such close proximity and the psychological trauma inherent not only to each killing act, but also to the many killing acts, would have been of such magnitude that no one would have let them happen.

As indicated on the earlier chart, killing conducted with unmanned systems belongs at the very end of the long-range killing spectrum and might even be worthy of having its own designation: killing at maximum range. It is reasonable to expect that at this range, killing is made extremely easy. The contention here is that there is no other tactical weapon currently on the battlefield that facilitates killing with such physical and psychological ease, such that killing becomes a rather clinical and dispassionate matter, easing any of their operators' existing moral qualms. Noel Sharkey offers support to this argument by drawing attention to reports collected by Singer.[74] Amongst a variety of other disturbing statements, he cites one 21-year-old soldier who talks about his acts of killing with casual indifference: 'the truth is, it wasn't all I thought it was cracked up to be. I mean, I thought killing somebody would be this life-changing experience. And then I did it, and I was like, "alright, whatever"'.[75] Later, he says that 'killing people is like squashing an ant. I mean, you kill somebody and it's like "alright, let's go get some pizza"'.[76] In this clinical killing environment, in which it seems reasonable to propose that some human targets are divested of their humanity when it comes to on-screen representation, respect for *jus in bello* norms is likely to wane. Here, some of the videos mentioned in the previous section can serve as a demonstration of this waning. Many show raw footage of Predator strikes which represent people as little more than 'hot spots' or 'blips' on the screen and which show operators often failing to take the necessary precautions to ensure noncombatants are protected.[77] Again, it seems that obscuring 'targets' in this way and increasing the distance to the maximum possible, makes it easier for unmanned systems operators to kill in an indiscriminate and disproportionate fashion.

While the distances involved[78] show the powerful role of unmanned systems in overcoming moral qualms and the socialised inhibition to killing, there is a range of other mechanisms that increase the distance between operators and their opponents, and which make it even easier for systems operators to kill without regard for the consequences of their actions or *jus in bello* norms. The first additional mechanism on which Grossman explicitly remarks – the one that is most relevant to America's use of unmanned platforms, but which can only be mentioned briefly here – is cultural difference. Indeed, it allows warfighters to further dehumanise the enemy by perpetuating racial and ethnic differences.[79]

Military forces have long been trying to get their troops to think of the enemy/ies as 'inferior forms of life … as less than human'.[80] Put simply, this is relevant in the drone case because the further one is technologically distanced from the enemy, the easier it is to think that they are significantly different, hence the easier it becomes to kill them. Unmanned systems separate soldiers from the cultural environment that they would operate in if they were in the field and, in that sense, permit them to racially demonise and alienate the enemy. Connected to cultural distance is moral distance, which involves legitimising oneself and one's cause.[81] Once it has been determined that the enemy is culturally inferior, it becomes less difficult for operators of unmanned platforms to incorrectly suppose that their enemy counterparts are either misguided in their pursuit of war or share their leaders' moral guilt. It also eases the thought that this warrants waging unconstrained violence against those with the supposedly morally inferior cause.[82] Strawser's argument, that the only combatants who can fight morally with unmanned systems are combatants who are objectively just, represents a similar sort of thinking. This will be the subject of critique in the following chapter.

However, not all are convinced by the argument that unmanned systems present a moral problem by making it easier to indiscriminately and disproportionately kill. Daniel Brunstetter and Megan Braun argue that these systems are subject to the same *jus in bello* requirements as other weapons used in war, but that their technological advantages coupled with the removal of risk to soldiers means that they should, at the least theoretically, more easily satisfy the principles of discrimination and proportionality, and perhaps make operators more reluctant to kill in situations where there is doubt with regard to the legitimacy of the potential victim of aggression.[83] They say that in the case of surveillance, at the very least, the distance or what they call 'separation factor', arguably offers an increased level of control over lethal targeting decisions and ought to actually reduce unnecessary killing and destruction.[84] They regard the fact that a drone operator can have the opportunity to confer with a superior officer as being a critical factor in encouraging ethical decision making in war. In some instances, this may be exactly right, but in others, it might be that having a mission commander overseeing the operator's actions only places additional pressure on the operators to fire (here bear in mind the sergeants walking the trenches of World War I, whose aim was to encourage reluctant soldiers to kill). Enemark also questions some of the assumptions relied on in this 'lower barriers' argument.[85] He says that there is reason to suppose that being physically absent from the battlefield is *more* conducive to discrimination. In his view, the removal of risk allows decisions to be made in a more deliberate manner by removing anger and emotion, which he thinks might otherwise lead to morally unsanctioned killings.[86] That is, if a drone operator working from a desk in Nevada encounters the enemy, adherence to *jus in bello* protocols should improve as the operator is at little or no personal risk. This may well be true in some instances and is obviously morally preferable. However, as will be argued below, if an operator is so emotionally removed, s/he is arguably just as likely to develop the sort disengagement referred to above or even a morally concerning callousness.

What needs to be recognised is that the disengagement and depersonalisation of war can go so far that operators begin to lose sight of the fact that they are actually involved in real wars. It is becoming increasingly difficult to distinguish whether one is playing a video game or operating a real unmanned system with weapons that will also have very real consequences for those at which they are directed. As Nolen Gertz notes, while it should be rather clear that video games try to approximate the reality of combat insomuch as it is possible for such a medium to do so – the attempts to do so have reached new levels when e.g. the designers of 'Medal of Honor' are turning to former members of the military and its special forces for technical guidance – what is probably a lesser known fact is that the military and its weapons designers often try to approximate the reality of war gaming as well.[87] Unmanned systems designers are currently trying to model the controls of these platforms to those of video game consoles or, in some cases, actually using the very same controls, wired especially for use with these systems. Singer cites a project manager who was responsible for designing an unmanned ground vehicle: 'we modelled the controller after the PlayStation because that's what these 18-, 19-year-old Marines have been playing with pretty much all of their lives'.[88] The argument is not just that these systems create a militaristic mentality; that would be too reductive. What is being said is that recruits, who are already familiar with the technology, have a premeditated and probably distorted idea of war. The related moral concern, of course, is that people often do things in video games that they would not do in real life and that this may somehow carry over to the use of unmanned systems. That is, they may be led to mistake reality for virtual reality, thus making them more prone to launching indiscriminate and disproportionate attacks, which are characteristic of the most violent video games. This further supports the argument that the increasing employment of unmanned systems tracks unethical conduct in war.[89] As will be argued in the next section, this inability to separate the two worlds may also have deep psychological effects on the operators themselves, only compounding the problems posed by distancing, disengagement, de-individualisation and the effect they have on adherence to *jus in bello* norms.

The Personal Wellbeing of the Distance Warfighter

In the previous section, it was argued that unmanned systems and the physical and psychological distances that they create magnify the challenges that warfighters face in applying and adhering to *jus in bello* norms. This section will be less focused on the mindset-altering effect or psychological impact that these systems have on the warfighting role of unmanned systems operators, and more focused on the impact on their *personal* wellbeing. This is relevant, given that at both the *jus ad bellum* and *jus in bello* level, the harm incurred to the initiating state's warfighters must be given some weight, even if it is not, or should not, be the overriding priority. Some suggest that killing via proxy does not have significant adverse psychological or emotional effects on the distance warfighter. Lambèr Royakkers, for instance, says that 'creating moral disengagement reduces or even eliminates the stress of cubicle warriors'.[90] Grossman also writes that in his years of reading and researching on the subject of killing, he has never come across a single instance of psychiatric or psychological trauma associated with such long-range killing.[91] Sharkey gives further anecdotal evidence supporting these claims in citing the commander of 432nd Wing at Creech Air Force Base, who said that 'on only four or five occasions had sensor operators gone to see the chaplain or their supervisors and that this was a very small proportion of the total number of remote operators'.[92] The reasoning employed seems to follow from an idea examined in a study conducted by Michael Osofsky, Albert Bandura and Philip Zimbardo.[93] They hypothesised that the selective disengagement that similarly distancing systems (or processes) facilitate, allow operators to maintain their emotional wellbeing, whilst performing their role, which involves killing people and conducting themselves in an injurious fashion. However, Sparrow expresses the concern that, while distancing unmanned systems operators from the risk of death and allowing them to become disengaged might be thought to reduce stress, killing by the push of a button can still be a stressful job and might actually expose operators to new, somewhat different or unexpected stressors.[94] What this reveals is that something is missing from the previous evaluations of what unmanned systems can reasonably achieve. It is particularly important to understand these new and/or different psychological stressors, because they will both influence moral decision making in war and guide public opinion as to whether unmanned warfare really is an improvement over earlier forms of warfare. In the latter sense of guiding public opinion, this discussion is relevant to that in the first section of this chapter.

Before delving deeper into the psychological issues that affect unmanned systems operators, we must comment more generally on the burden of killing as it affects conventional warfighters in regular close and midrange warfare, primarily to provide a comparison point. As has already been mentioned, the resistance to killing at these distances can often be so extreme that it overcomes not only one's own innate self-preservation mechanism, but also the coercive force exerted by sergeants and platoon commanders. Moreover, it can override the expectations of one's peers and the more general obligation outlined in Chapter 3, i.e. that of defending the state and its otherwise vulnerable population. As such, the ordinary warfighter is put in a situation where they cannot avoid some sort of psychological trauma because of conflicting feelings, stresses and moral obligations reinforced by society, by a range of regulations and procedures that govern wartime conduct. It has been claimed that, if the conventional soldier can overcome the socialised resistance to harm and kill individuals in close or midrange combat, he will forever be burdened by guilt. A suitable example is given by William Manchester, a novelist and United States Marine veteran of World War II, who describes his feelings after killing a sniper in a fishing shack who was picking off the Marines in his company, one by one: 'I shot him with a 45 and I felt remorse and shame. I can remember whispering foolishly, "I'm sorry" and then just throwing up … I threw up all over myself. It was a betrayal of what I'd been taught since a child'.[95] Yet, if for some reason a soldier does not kill, he has also failed to fulfil his duty to his comrades. John Early, a former mercenary and Vietnam veteran, explains that 'you're depending on him [the man next to you] for the most important thing you have, your life, and if he lets you down you're either maimed or killed. If you make a mistake, the same thing happens to him, so the bond of trust goes extremely close'.[96] Furthermore, the reluctant soldier is subject to the shame that comes with failing his profession, fellow citizens and state. In Grossman's words, the soldier is 'damned if he does, and damned if he doesn't'.[97]

For conventional soldiers, it becomes a matter of balancing, on the one hand, the obligation to do what is militarily required in order to protect one's nation, which is generated by the society that one serves and the comrades with whom one serves alongside, and on the other hand, the guilt that inevitably follows killing in war. From their comments above, Royakkers and others seem to imply that due to the moral buffers introduced by unmanned systems, this balancing act is made significantly easier for the soldiers-turned-unmanned systems operators. If we follow Grossman's line of thought, we might even conclude that this balancing act is totally insignificant or not even a factor for these distance warriors. To make a comparison with combat at close or midrange, it might be said that the psychological stress profile of a drone operator is much more similar to that of a higher commander, who may have some operational function, but who is much more strategically focused. Unlike the average soldier, but quite like an unmanned systems operator, high level military decision makers generally do not face the same sort of risk that burdens the troops that are physically present at the frontline.[98] Most of the time, these high level decision makers go to sleep at night knowing that they are very unlikely to be bombed or mortared during their sleep; they do not have to worry that field communication lines might be cut or that resupply might not happen; and they generally are not faced with the task of retrieving the dog tags of those who have been killed in action beside them or writing personal letters home to their loved ones.[99]

While this may well be true, both of higher commanders and unmanned systems operators, the more relevant similarity is arguably that they both face a range of other psychological stresses that, while qualitatively different, seem to equal or surmount those faced by conventional combatants and combatant commanders. For instance, higher commanders may not have to worry about being killed by the enemy, but they too have to be concerned about the enemy's movements and how they impact operations.[100] They may not be worried about impediments to their own progress or dangers to their own life, but they are likely to be worried about ensuring that their troops have sufficient resources to conduct their campaign, their commanders' ability to conduct operations and the views of the public and the media.[101] Likewise, unmanned systems operators face their own unique problems. While some remote operators may be spared from the raw first-hand experience of seeing people die on the battlefield, in many ways, they may actually see more of war and its deadly outcomes than any typical soldier engaged in close or midrange warfare. There are two key reasons for this, one more convincing than the other. Firstly, while it has been argued that most of today's unmanned systems reduce targets to a 'blip' on the screen, many say that some systems have such high-resolution optics that operators can see the impact of their actions with greater clarity than most conventional air force pilots.[102] Again, there is doubt about this because of the physical limitations with which optical technologies must operate, but technological developments may make this more of a concern as time passes. However, unmanned systems do improve loiter capability over conventional manned systems. This means that an operator may observe a target for 12 or more hours before taking lethal action.[103] Thus, even if the raw footage is not particularly clear and the operator is somewhat disengaged, this extended surveillance period, in which many disturbing events may unfold, has the potential to be psychologically damaging. This damage can happen, if not at the time of operation, at a later stage when the operator has had time to reflect on what has happened and what role s/he has played in the events. This is the first sort of morally relevant stress we will be concerned about: that of witnessing traumatic events. It is not distinct to unmanned warfare, but its implications are unique and significant when considered in relation to other compounding factors.

Another associated stress rarely mentioned in the academic literature is that related to limited participation in these events. Like a higher commander who has to exert influence on the ground via troops of which he is not directly in control, and with only information conveyed by field-based combatant commanders, a drone operator has to contend with having a limited capacity to intervene or influence events that are unfolding in the combat zone. Sparrow recognises that this general concern is more significant than it perhaps seems because, despite the 'hype' surrounding them, many unmanned systems are limited in the information they can relay, the force they can actually exert and the range of actions that they make possible.[104] As mentioned in the previous chapter, it is not always the limitations of the systems themselves, but often the capacity of these systems to be utilised within particular environmental surroundings which limits adherence to just war theory and creates psychological problems. In his first-hand account of piloting drones, Matt Martin details the anguish he experienced resulting from being unable to shape or rectify some injustices he witnessed on the battlefield.[105] He described how he had carefully planned to kill a group of enemy combatants and recounted the feelings he experienced when two children walked into the attack path post-missile release. '[T]here was nothing I could do to save them … we could only stare in abject horror as the silent missile bore down upon them'.[106] The feeling of helplessness and shame that Martin tells of is compounded when it is the enemy waging the violence and one's own comrades who stand to be injured or killed. In much the same way as a missile release gone badly, an unmanned system may limit the distance warrior's moral agency by limiting their capacity to act in reference to right and wrong. In a combat scenario, this means they cannot aid fellow warfighters in the same way they could if they were physically present on the battlefield.[107] Benjamin cites a young sensor operator who reported, having witnessed the death of troops from afar, that 'it was a traumatic experience. It shocked [him]'.[108] The dissonance that comes from sitting in a peaceful country in relative safety and taking indirect action, while seeing fellow military personnel die, is one that is unique to unmanned systems operators and one that operators need time to process and understand.

Unfortunately, this time for processing potentially traumatic experiences, or lack thereof, is another problem with consequences for ethical conduct in warfare. Unmanned systems operators are under ever increasing operational pressure driven by cost cutting and unrealistic expectations. A recent US Air Force study revealed that one of the biggest residual sources of stress for drone operators were the long hours and frequent shift changes that occurred due to staff shortages.[109] These staff shortages could be seen as offering further anecdotal evidence that there is a serious concern about the stress to which these operators are exposed. Often-intense working hours and constant shift changes deny an operator time to critically reflect on their actions and prevents them from fully integrating into ordinary civilian life once they leave the base. Unmanned systems have fundamentally changed what the experience of 'going to war' is like. As Singer writes, it has given war a whole new meaning. For the last 5,000 years, when we were describing somebody as going to war in defence of the state – whether it be the ancient Greeks going to war against Troy or coalition forces fighting against the Iraqis in the Gulf War – we were talking about them physically going to the battlefield, putting themselves in danger, knowing that they may never come home or see their family again.[110] That is what 'going to war' has involved for the last five millennia, but the experience of war has changed.

One US Air Force Pilot who Singer interviewed for his book describes how he would wake up in the morning, drive to the base and then spend 12 hours putting 'missiles on targets' (note the dehumanising language) before driving home again. Twenty minutes after he had been at war, 'he would be at the dinner table talking to his kids about their school work'.[111] However, this does not mean that modern unmanned warfighters are at less psychological risk than those physically present in the battlespace. This new experience of being at war – of being simultaneously at home and at war and having to balance the two – is itself a major source of stress that distance warfighters encounter every working day. As we are only in the early stages of understanding this problem, few studies detail the psychological effects of this juxtaposition between being at peace and being at war, but Martin hints at the significance of this problem in a powerful example given on the final page of his book.[112] Having just killed a number of enemy combatants and destroying the building they were housed in, he recalls how on the way home to see his wife he pulled his 'car into an In-and-Out Burger, changed out of uniform, and grabbed a burger and fries. Only after I was standing in line … did I have time to reflect on what had just happened, how unreal this war truly was'.[113] This and the mounting civilian death toll seem to highlight that, while there may be much that we do not know, the human warfighter may not yet be capable of effectively handling the dual experience of killing people at war and being at peace, coping with the domestic demands of children, partners and society. Going home after a day of combat does not necessarily afford an operator time for reflection, for few will go home and talk over the dinner table about the combatants or noncombatants that they have killed that day, no matter how righteous the killing. Instead, they keep it to themselves and must then find some way of dealing with the constant disjunct between military and civilian life.

In many cases, it may prove morally preferable to expose operators to these unique stresses rather than those they would experience if they were in the field themselves.[114] However, others seem to view it is as another eliminable problem concerning distance and that, just as we have removed the warfighter from the battlefront, we should now remove the warfighter from the homefront.[115] Subscribing to this sort of thinking, one military commentator has put forward measures that he thinks will maximise efficiency while alleviating the military-civilian disjunct.[116] These include establishing a 'communications bubble' that would see all personal phone calls to and from the control room banned, keeping units sequestered at a hotel or at barracks on a base before and after any large operation, so as to isolate from their families and civilian lives in the hope of eliminating the accompanying troubles.[117] This may seem good operational thinking but, if anything, it merely postpones the eventual meeting of the two opposing worlds (civilian and military) and may introduce further problems affecting ethical decision making if operators come to feel unnecessarily forced apart from their families and civilian lives. The attention is best focused elsewhere. It should be realised, for instance, that distance warfighters do not benefit from the same sense of camaraderie that conventional warfighters do.[118] A squadron of drone operators is rarely together and never has the time to invest in the shared experience of the group or simply rest and recover. It would prove better to treat the stress more as a manageable rather than eliminable problem, building camaraderie or otherwise offering services that assist with management of the psychological stressors that come with unmanned warfare. This may more successfully lead to improving operators' sense of wellbeing and ethical conduct in warfare.

Having this sort of support mechanism in place is particularly important in unmanned warfare as studies have revealed that the proportion of inexperienced to experienced personnel is much greater than in conventional warfare and the former may not be able to cope with the emotional demands of unmanned warfighting on their own.[119] Drone operators may be attracted to rendering military service in this form because they are familiar with the controls and the gaming experience that it replicates, or simply because it seems less risky and allows them to maintain a connection with loved ones they are reluctant to leave behind. However, many presumably do so in complete ignorance that, by fighting from the home front, they actually cut themselves off from the support and comradeship which one finds in a typical ground unit. The perception of the distance warrior is another factor that must be taken into account, for this has bearing on the level of support that one receives from both within the military and the public. The image of the 'top gun' combat pilot has been portrayed in novels and films and it is this image that likely resonates with the military leadership and the public. Unmanned systems operators do not seem to receive the same level of gratitude for rendering public service as these conventional pilots. Neither do these operators get the welcoming home parades or medals. The American Distinguished Flying Cross, for instance, has never been awarded to an unmanned aircraft pilot and many think this is for good cause.[120] These commendations tend to be given to reward bravery and physical risk. Hence unmanned system operators do not receive them because they are not perceived as putting themselves at such risk. Their machines take the risk, so to speak. Until other forms of risk come to be seen as significant enough to warrant reward, it seems important for distance warfighters to initially establish support groups amongst themselves, particularly because our current understanding of the psychological attributes considered critical to performing well in unmanned warfare and coping with the results is at such an early stage.[121]

All of the aspects of combat stress and trauma discussed here have the potential to profoundly impact not only the wellbeing of the warfighter, but the individual's contribution to the war effort and the overall contribution of the aggregates of individuals that we call squadrons and military units. If we begin to understand the range of responses had by unmanned systems operators, we can begin to understand how mediating warfighting technologies, such as drones, influence ethical conduct in warfare and then seek to make the necessary changes. As already mentioned, we do not yet have the psychological profiles that are necessary to select the most psychologically well-suited candidates for unmanned warfare and, even then, there is no guarantee that this will lead to more discriminate and proportional warfare. We also have a situation whereby political and military leaders may utilise drones to alter the public mindset vis-à-vis engagement in war and to lower barriers to killing. For these reasons, many have suggested that we focus on technical solutions to what is a technologically mediated but largely human problem. The discussion concerning the responsibility of system engineers and designers will be developed in Chapter 9, but some of their proposed technical solutions are described below.

Technological Solutions to this Human Problem?

In the previous two sections it was said that in practice, human warfighters often fail to discriminate between legitimate and illegitimate targets for at least two reasons. First, killing at maximum range temporarily reduces the humanity of one's enemy and detaches the distance warfighter from the consequences of the use of their weaponry. Second, this often results in traumatising psychological consequences for those warfighters when they eventually come to grasp their actions, which may subsequently lead to further unethical conduct in wartime. Unfortunately, when it comes to trends in adherence to legal and ethical norms, the human record in adhering to the dictates of *jus in bello* principles is questionable at best. In fact, Arkin dedicates an entire chapter in his book *Governing Lethal Behavior in Autonomous Robots* to human failings on the battlefield.[122] He cites the US Surgeon General's report into battlefield ethics and the mental health of soldiers deployed in Operation Iraqi Freedom, which details some concerning findings.[123] Among them were claims that: approximately 10 per cent of soldiers reported mistreating noncombatants; less than 50 per cent agreed that noncombatants should be treated with dignity and respect; roughly 15 per cent agreed that all noncombatants should be treated as insurgents; and close to 30 per cent reported facing ethical situations in which they did not know how to respond.[124] Arkin suggests that with an increase in the use of technology like semi-autonomous drones, the figures noted in these reports will become even more concerning and that atrocities in war will also become increasingly worse. His primary conclusion is that, because of their fallible nature, human warfighters cannot be expected to adhere to wartime standards when they are confronted by the horrors of the battlefield and we ought to develop a technology that extends beyond the limits of human warfighters.[125] Arkin's own solution is to eliminate the warfighter from the decision loop altogether, but this is technologically infeasible as things stand. There are, however, a number of other options with varying levels of autonomy that will first be investigated. All are important because they are seen as solutions to the potential for unethical decision making discussed in the preceding sections and also because the focus in the next chapter will be on radically asymmetric conflict and semi-autonomous or autonomous technology.

Mary L. Cummings was one of the first engineers to recognise that, because of the inherent complexity of socio-technical systems such as unmanned systems and the decision support systems with which they are integrated, users may perceive the computerised system to have some form of legitimate authority, diminishing moral responsibility and shifting accountability, thus creating a moral buffering effect similar to that described earlier.[126] For her, the solution was not necessarily to invest in machines with high levels of autonomy, but for engineers and system designers to pay particular attention to the weapons control interfaces of non-autonomous and semi-autonomous systems and how they influence the potentially lethal decisions of their human operators.[127] Cummings suggests that a methodology designed in human-machine interaction research, the value sensitive design approach (VSD), can bridge the technical/operational requirements of the military and the moral requirements of theories such as just war. The idea being that human values should be at the forefront of technological design. As a case study, Cummings applies the VSD approach to the design of a supervisory command and control system of a Tomahawk Missile. Upon feeding research participants incorrect information and challenging them to evaluate their own design profiles in relation to pre-specified moral values, it was found that operators without a decision aid outperformed those with a decision aid. This reveals that the interface design directly impacted targeting decisions and that there appeared to be a tradeoff between effectiveness and efficiency (operational values) and the safety of noncombatants (a moral value).[128] However, while this sort of approach may be useful in training engineers responsible for the design of unmanned systems, it merely examines an operator's performance against pre-identified values. It provides no method to weigh these values, as is required by just war theory, nor does it provide an explanation as to how the weapons control interface influences targeting decisions in any systematic way. Therefore, it is not an all-encompassing solution to unethical conduct in unmanned warfare.

Asaro argues that what is needed in the technical design process is an understanding of what kind of information people use in making value-laden lethal targeting decisions, how people process this information and, mostly importantly, how various ways of presenting and representing that information directly or indirectly influence ethical decision making.[129] He points out that this kind of psychophysical analysis has been done in the past by the military and was also common in the field of 'human factors' during World War II, due to the reliance on humans in critical targeting tasks.[130] Asaro faces an additional challenge, however, since these methods have only be used after the fact, meaning that they had performance criteria against which to judge. To go about understanding how users perform in making real-time value-laden targeting decisions, where performance is much harder to judge, he proposes that we start by modelling ethical decision makers. His 'modelling of the moral user' involves three key elements.[131] First, we would need to draw on cognitive psychology to understand the representations, decision rules and emotional requirements for proper ethical decision making. He says that this might involve undertaking detailed empirical studies of some of the emotions mentioned in the previous sections: fear, anger, sympathy and so on. Second, we would need to use recent work in experimental psychology on moral intuition, value comparisons, and call on experimental economics to understand the nature of risk assessment and probability estimation. All this would be needed in order to get some understanding of what sort of framework people use for their ethical decision making, while keeping in mind that many do not use any kind of recognisable framework. Third, we would need to determine what standards we, as a society, want to hold soldiers to, to what extent we can impose these on soldiers through technology and, further, what level of psychological risk it is preferable to expose our warfighters to.[132] While this approach seems to be better than Cummings's, because it can potentially consider more complex ethical problems with systems before they are actually fielded, his proposed system of modelling the imperfect human warfighter may not be that much better than the actual imperfect human warfighter. That is, why go to such great effort and cost when the whole idea is to leave behind the problems posed by having humans in the loop? Asaro himself seems to agree and he suggests that, at the very least, we might recognise various psychological contradictions and begin to understand the range of diverse ethical considerations that end-users employ in their role as the ultimate arbiters of just war theory.[133] Indeed, this would be most valuable, but others would prefer to skip this step.

For Arkin, the solution is not to tailor the weapons control interface to encourage ethical decision making in war, but rather to eliminate the source of unethical decision making by removing humans and the human-machine interface altogether, and to look at how one might actually provide unmanned systems with the capacity to make ethical decisions and carry out different kinds of actions, some of them lethal. He is convinced that ethical programming is possible and that it would not only meet the normative requirements of international humanitarian law and just war theory, but would also ultimately surpass their requirements. He writes: 'I am convinced that [unmanned systems] can perform more ethically than human soldiers are capable of'.[134] Arkin proposes that we equip unmanned systems with an 'ethical governor', which is based on Watt's mechanical governor for the steam engine.[135] This governor acts as a transformer or suppressor of automatically generated lethal action. So, in much the same way that the mechanical governor would shut down a steam engine running too hot, the ethical governor shuts down the autonomous machine when it is about to do something unethical, such as firing too closely to civilians (the parameters for which would have to be set by humans, at least in the first instance). But this is just one component. The second component is the 'ethical behavioural control' unit, which monitors all of the systems behaviours and ensures that they evolve from within set constraints.[136] The third is the 'ethical adaptor'. It allows the system to learn, adapt and overcome by updating the system's set of constraints.[137] It is presumably intended to act in place of human guilt or remorse in case some unethical conduct manages to pass the previous two components. The final component is the responsibility advisor, which ensures that, in the event of a lethal action occurring, responsibility can be clearly allocated to the human or the machine, in the case of joint interaction.[138]

Arkin believes that with these sub-systems working effectively together, autonomous machines would possess the capability to operate morally, ensuring that any force is delivered proportionally and with maximum discrimination between combatant and noncombatant targets. However, he does not seem to recognise that there is a difference between a machine that acts in accordance with the dictates of morality and what Wendell Wallach and Colin Allen call a 'moral machine'.[139] That is, he fails to recognise that there is a distinct difference between following moral rules or laws and actually being moral in the full and proper sense of the word. While the technical acts of discriminating between targets and determining proportionality may be a necessary condition for moral conduct in war, they are not sufficient. For instance, Andreas Matthias has said that Arkin seems to view discrimination as a mere matter of technical classification between friend and foe and between legitimate and illegitimate targets of potentially lethal military action.[140] He reduces discrimination to a cognitive level operation that the machine mimics. To him, all the machine does and needs to do is to identify all the targets within its specified radius of operation and to sort them into various categories: combatants and noncombatants, children, wounded persons and so on. While this is an important capacity for any moral agent in war, Arkin's system does not actually establish what these categories are, it presupposes them. 'Individual X standing beside weapon Y is Z' (an instance of discrimination), is a straightforward classification result and there is no sort of moral deliberation involved. To ask whether it is morally right for the machine to kill individual X standing next to weapon Y, on the other hand, involves genuine moral evaluation. For this very reason, if no other, Arkin is incorrect in saying that his proposed system can outperform human warfighters.

Part of the problem is that Arkin takes the laws of armed conflict, the rules of engagement and the just war theory to be mutually compatible and translatable into a set of rules that are also compatible with his control architecture and programming language. When the principles of just war theory are viewed in this way, it is easy to mistakenly think that what is left is a technical matter of ensuring that the machinery is capable of correctly and reliably carrying them out. In reality, although many military and political leaders also portray these sources of restraint as if they were a clear and steadfast set of rules amenable to automation, each is actually a set of highly context-dependent guidelines. The law of armed conflict is, in Asaro's words, a 'menagerie of international laws, agreements and treaties' that are open to considerable interpretation.[141] Rules of engagement, which are devised to instruct warfighters how to act in specific situations, can also be rather vague, as they are an attempt to combine legal, political and strategic concerns on one small card and leave much up to the individual soldier. Likewise, just war theory is often presented as an easily applicable theory, when in most formulations it is, by its very nature, a set of principles that requires a great deal of moral deliberation for the values and rules of thumb it sets up to be put to effective use. Take the *jus in bello* proportionality principle, for instance. There is a host of difficulties in operationalising this principle and in carrying out the necessary calculations in the field. As Lucas says, we do not have anything like a 'formal proportionality algorithm' and it requires weighing and comparing incommensurable goods under conditions of extreme epistemic uncertainty.[142] How many men, women and children, for example, is it permissible to kill for a given military objective? Questions like these require a kind of moral evaluation of which Arkin's system is incapable. He might advance the more limited claim that his autonomous drones might outperform humans in classificatory tasks, but he wants to make the stronger but less plausible claim that his proposed system will perform more ethically than humans.

Even if we put aside the fact that he glosses over the difference between representing morality and acting morally, both Lucas[143] and Sparrow[144] highlight that there is an additional conceptual problem associated with saying that an autonomous unmanned system might perform better than a human warfighter. As has already been said, one ought to be willing to concede that Arkin's proposed system might function better in classificatory tasks and thus might perform better under certain limited wartime conditions. We might discover through an empirical study, for example, that the failure rate of human warfighters confronting some specific sort of decision-making scenario is 10 per cent or, alternatively, that they comply with the necessary legal and moral restraints 90 per cent of the time. In this instance, it means that the human warfighters are fallible. Suppose that we can reliably track compliance with the relevant rules and principles. Arkin simply demands that his system perform as well or better than this 90 per cent benchmark set by the human warfighter. When we send a human warfighter into battle, we hope for 100 per cent compliance with the laws of war and just war theory, even though we know all too well that, for the reasons stated earlier, soldiers are often statistically unlikely to meet these standards. But when human warfighters fail to do so, they are held accountable and punished. The problem with Arkin's argument is that he is effectively lowering the moral benchmark for conduct in war, which may result in further breaches of the law of armed conflict as well as the proportionality and discrimination principles. If we know in advance that, in a particular scenario, Arkin's system will perform with less than 100 per cent compliance with the relevant laws and moral principles, it would be unethical to have it operate.

The solution, while not endorsed here, may be to go beyond the sort of robotic architecture that Arkin proposes and to develop systems that truly have full autonomy or what others call 'strong artificial intelligence'. That is, it might be morally preferable to develop a system that can create and complete its own tasks without the need for any human input (except, of course, for the decision to build itself) instead of Arkin's system which still calls on a human to make certain decisions when it cannot.[145] As per the definitions provided in Chapter 1, the human is so far removed from the loop that the level of direct influence on this sort of system is totally negligible. Such a system would display capacities that imitate or truly replicate the moral capacities of sentient human warfighters and would therefore be able to engage in the sort of moral deliberation that the laws of war and just war theory demand. There is considerable disagreement in the disciplines of artificial intelligence and philosophy regarding the point at which we can reasonably expect machines to become as capable as human warfighters on the battlefield or whether this 'strong artificial intelligence' is at all possible. In 1965, Gordon Moore, Chairman Emeritus of Intel Corporation, described 'Moore's law', which predicted the doubling of transistors on silicon chips with each new generation leading to rapid acceleration in the power of computer, and it has held true to date.[146] With this in mind, futurist Ray Kurzweil has prophesised that we will reach a 'technological singularity' – the point at which machine intelligence will surpass human intelligence – by 2045.[147] Some advanced the more reserved and plausible claim that, while computers may exceed human intelligence in a few rather narrow domains, they will be never be as universally intelligent as is necessary for moral conduct in war. At the other end of the spectrum, there are those who hold no hope. Joseph Weizenbaum has famously stated that 'no other organism, and certainly no computer, can be made to confront human problems in human terms'.[148] The philosopher John Searle has also argued that a computer may be able to solve a complex problem, such as those encountered in war, but it would lack any actual understanding of them.[149]

Despite the fact that Weizenbaum and Searle seem to dismiss the idea of fully autonomous unmanned systems as the stuff of science fiction, it is wise to briefly consider whether they could even be a solution to immoral conduct both in resorting to war and conducting war. Indeed, a great deal of money is being invested in research which works toward developing fully autonomous unmanned systems, and it seems impossible to rule out that they might be created in the future. To be sure, the potential problems are many. One of the immediate concerns that surfaces about waging warfare with artificial moral agents is that they may eventually turn on human beings. Again, this might sound fantastical or highly unrealistic, but even if we want to endow our unmanned systems with full moral agency (this cannot be assumed), this will involve giving them the capacity to contemplate both moral and immoral actions and it is a matter of conjecture as to what sorts of decisions autonomous military systems would make. A robot warfighter may not be motivated by power or dominance, as human warfighters are, and it may not have the same natural killer instinct, but they too may come to develop what some might claim are needs and aspirations, for which they will need to compete with other agents.[150] There is no guarantee that these systems will not evolve into the functional equivalent of today's unethical warfighters. What is more worrying is that these systems and the moral decisions they make may become so complex that humans will be incapable of understanding them. Imagine if these systems eventually become so powerful that we come to cede power to one super system and entrust it to make *jus ad bellum* decisions in lieu of human political and military leaders, which often deceive the public. The variables this super system would take into consideration might be so numerous that no human could hold all of them in his/her mind and understand whether a particular war was just or unjust. Strawser argues that with machines filling the decision-making niche that humans once did, it could eventually get to a point, where humans' individual moral choices are no longer needed or are rendered so insignificant that it would be better for them to surrender their moral agency altogether.[151] Under his account, both war and human life could be made meaningless. Of course, while this is logically conceivable, most engineers suggest that this is highly improbable and unworthy of a closer examination. However, such concerns are a reminder that we must critically examine the long-term implications of removing humans from the decision-making loop and the extent to which this is morally desirable.

Each of the technological solutions discussed above has a problem: some do not provide us with the insight necessary to weigh values and design ethical weapons control interfaces; some conflate the difference between acting morally and technical acts of discrimination and proportionality; and some might use their capacities to undertake actions which are neither desired nor capable of being understood by human beings. From a moral standpoint, the technological challenge is to find an appropriate trade-off between military effectiveness and harm to warfighters and human beings in general. Moreover, we must determine what elements of these solutions are worth developing. It seems that Asaro's modelling of the moral user will be useful in semi-autonomous systems, that there is a limited role for Arkin's classificatory system, and that there is some use for autonomy in more repetitive warfighting functions. But none of these solutions alone is sufficient for overcoming the mindset-altering effect, psychological challenges and unethical conduct linked to unmanned warfare. That is, the solution to this technologically induced human problem does not necessarily require the introduction of more technology.

Conclusion

What this chapter has shown is that, alongside the more technical issues that plague unmanned systems, there is a mindset-altering effect which has a significant impact on the ethical compass of our principal war-making agents, from unmanned systems operators through to political and military leaders and the public at large. In the first section, it was argued that the risk transference discussed in the previous chapter, along with some linguistic subterfuge and media management, results in a cognitive dissonance for the public and a lowering of the right authority threshold, thus making it easier for high-level decision makers to take an unwilling public to war. It was suggested that we should strengthen the right authority requirement to guard against this. In the second section, it was demonstrated that we must also be concerned with how technologically mediated fighting lowers barriers to conduct in war. It started by outlining the various killing ranges and locating unmanned systems at 'maximum range', where killing is facilitated by physical and psychological distance, which leads to a dehumanisation and deindividualisation of the enemy. This section concluded that this aspect, coupled with the virtual nature of unmanned warfare, makes it significantly easier for soldiers to kill without proper respect for the *jus in bello* principles of discrimination and proportionality. In the penultimate section, it was argued that, despite some arguments to the contrary, the constant disjuncture between being at war and being at peace, and having merely a few psychological stress management tools, further magnifies the challenges of adhering to *jus in bello* norms. The final section examined some technological solutions to this largely human problem. It argued that we really need humans to make some wartime decisions, and that we must combine our best efforts in designing weapons control interfaces, as well as judiciously introduce autonomy, to ensure that support for the aims of just war theory does not wane. In the next chapter, we will look at a second formulation of the threshold problem pertaining to asymmetry, a problem that falls at the intersection of *jus ad bellum* and *jus in bello.*

Notes

1 George Lucas, 'Industrial Challenges of Military Robotics', in *The Ethics of Emerging Military Technologies* (University of San Diego: International Society for Military Ethics, 2011); Lucas, 'Postmodern War'; Edward Barrett, 'Executive Summary and Command Brief', *Journal of Military Ethics* 9, no. 4 (2010); George Lucas, 'Industrial Challenges of Military Robotics', *Journal of Military Ethics* 10, no. 4 (2011).

2 Robert Sparrow, 'Killer Robots', *Journal of Applied Philosophy* 24, no. 1 (2007); Robert Sparrow, 'Predators or Plowshares? Arms Control of Robotic Weapons', *IEEE Technology and Society* 28, no. 1 (2009).

3 Singer, *Wired for War: The Robotics Revolution and Conflict in the 21st Century.*

4 Lin, Bekey and Abney, 'Autonomous Military Robotics: Risk, Ethics, and Design'.

5 Lucas, 'Industrial Challenges of Military Robotics', 276.

6 Ibid.

7 Asaro, 'How Just Could a Robot War Be?', 55.

8 Ignatieff, *Virtual War: Kosovo and Beyond*, 176.

9 Ibid.

10 Ibid., 177.

11 This problem was probably also at play in early 2011, as the United States, along with some European and Arab states, debated whether they should become involved in maintaining a no-fly zone over Libya to combat Gaddafi's oppressive rule over its people. Recent events in Syria have raised the same question. For more, see David Whetham, 'Remote Killing and Drive-by Wars', in *Protecting Civilians During Violent Conflict: Theoretical and Practical Issues for the 21st Century*, ed. David W. Lovell and Igor Primoratz (Farnham: Ashgate, 2012), 203.

12 Singer, *Wired for War: The Robotics Revolution and Conflict in the 21st Century*, 323.

13 Zack Beauchamp and Julian Savulescu, 'Robot Guardians: Teleoperated Combat Vehicles in Humanitarian Military Intervention', in *Killing by Remote Control: The Ethics of an Unmanned Military*, ed. Bradley Jay Strawser (New York: Oxford, 2013), 122–23.

14 In constitutional democracies, it might be that the declaration of war actually calls for a non-constitutional referendum or a plebiscite, in which all affected persons (usually an entire electorate or country) are required (or at least eligible) to vote for their preferred course of action. A majority rule is required for any given action, a war or otherwise, to be approved.

15 Ignatieff, *Virtual War: Kosovo and Beyond*, 177.

16 Ibid.

17 Anthony Cordesman, 'The Lessons and Non-Lessons of the Air and Missile War in Kosovo' (Washington, DC: Center for Strategic and International Studies, 1999), 9.

18 Ignatieff, *Virtual War: Kosovo and Beyond*, 177.

19 In fact, the British Ministry of Defence has even commissioned a study in which it is recommended that contractors employ drones rather than military personnel in an attempt to distort the public's perception of the risks associated with war. See Ben Quinn, 'MOD Study Sets out How to Sell Wars to the Public', *The Guardian* (2013), http://www.theguardian.com/uk-news/2013/sep/26/mod-study-sell-wars-public.

20 Phillip Knightley, *The First Casualty: The War Correspondent as Hero and Myth-Maker from the Crimea to Iraq* (Baltimore: Johns Hopkins University Press, 2004).

21 Kenneth Payne, 'The Media as an Instrument of War', *Parameters: The US Army War College Quarterly Senior Professional Journal* 35, no. 1 (2005): 84.

22 Ibid., 84–8.

23 Keith Shurtleff, 'The Effects of Technology on Our Humanity', *Parameters: The US Army War College Quarterly Senior Professional Journal* 32, no. 2 (2002): 104.

24 Benjamin, *Drone Warfare: Killing by Remote Control*, 154.

25 Michael Strangelove, *Watching YouTube: Extraordinary Videos by Ordinary People* (Ontario: University of Toronto Press, 2010), 153.

26 Ibid.

27 Ibid.

28 Benjamin, *Drone Warfare: Killing by Remote Control*, 155.

29 Ibid.

30 YouTube, 'UAV Kills 6 Heavily Armed Criminals', http://www.youtube.com/watch?v=gNNJJrcIa7A&list=PL5FC6E7FB6B2FA591&index=4&feature=plpp_video.

31 Ibid.

32 YouTube, 'Hell Is Coming for Breakfast', http://www.youtube.com/watch?v=bHCchnGdtJA; Benjamin, *Drone Warfare: Killing by Remote Control*, 155.

33 Benjamin, *Drone Warfare: Killing by Remote Control*, 155.

34 Whetham, 'Remote Killing and Drive-by Wars', 203.

35 David J. Betz and Sangho Lee, 'Information in the Western Way of Warfare – Too Much of a Good Thing?', *Pacific Focus* 21, no. 2 (2006): 220–21.

36 Ibid.

37 Strawser, 'Moral Predators: The Duty to Employ Uninhabited Aerial Vehicles', 359.

38 James K. Bryant, *The Battle of Fredericksburg: We Cannot Escape History* (Charleston: The History Press, 2010). See opening to Chapter 6.

39 Patrick Lin, 'Ethical Blowback from Emerging Technologies', *Journal of Military Ethics* 9, no. 4 (2010): 325; Lin, Bekey and Abney, 'Autonomous Military Robotics: Risk, Ethics, and Design', 75.

40 Lin, 'Ethical Blowback from Emerging Technologies', 75.

41 Strawser, 'Moral Predators: The Duty to Employ Uninhabited Aerial Vehicles', 360; Lin, 'Ethical Blowback from Emerging Technologies', 75.

42 Strawser, 'Moral Predators: The Duty to Employ Uninhabited Aerial Vehicles', 359.

43 Ibid.

44 Ibid., 360.

45 Beauchamp and Savulescu, 'Robot Guardians: Teleoperated Combat Vehicles in Humanitarian Military Intervention', 106.

46 Ibid., 109.

47 Daren Bowyer, 'Just War Doctrine: Relevance and Challenges in the 21st Century' (Doctor of Philosophy, Cranfield University, 2008), 251–3.

48 A particular humanitarian intervention could also be supererogatory, i.e. may not be morally required of third party states, but nevertheless a good thing to do. Alternatively, it could turn out to be neither obligatory nor supererogatory. But this alternative does not tend to be taken seriously by the literature on the topic of the responsibility to protect the peoples or states that cannot protect themselves. Although this literature is still evolving, Savulescu and Beauchamp note that most of it is divided along 'obligatory-versus-supererogatory' lines.

49 Beauchamp and Savulescu, 'Robot Guardians: Teleoperated Combat Vehicles in Humanitarian Military Intervention', 117.

50 Michael Ignatieff, *The Warrior's Honour: Ethnic War and the Modern Conscience* (London: Vintage, 1998), 118.

51 Bowyer, 'Just War Doctrine: Relevance and Challenges in the 21st Century', 276.

52 David Grossman, *On Killing: The Psychological Cost of Learning to Kill in War and Society* (Boston: Little, Brown and Company, 1995).

53 Ibid., xiv.

54 Samuel Lyman Atwood Marshall, *Men against Fire: The Problem of Battle Command* (Norman: University of Oklahoma Press, 2000), 54.

55 Ibid., 71, 78.

56 Robert Emmet Meagher, *Herakles Gone Mad: Rethinking Heroism in an Age of Endless War* (Northampton: Olive Branch Press, 2006), 12.

57 Grossman, *On Killing: The Psychological Cost of Learning to Kill in War and Society*, 4.

58 Marshall, *Men against Fire: The Problem of Battle Command*, 4.

59 Grossman, *On Killing: The Psychological Cost of Learning to Kill in War and Society*, 97.

60 Glenn J. Gray, *The Warriors: Reflections on Men in Battle* (New York: Harper & Row, 1959), 178.

61 Ibid.

62 Grossman, *On Killing: The Psychological Cost of Learning to Kill in War and Society*, 98.

63 John Keegan and Richard Holmes, *Soldiers: A History of Men in Battle* (New York: Viking, 1986), 266.

64 Grossman, *On Killing: The Psychological Cost of Learning to Kill in War and Society*, 115.

65 Keegan and Holmes, *Soldiers: A History of Men in Battle*, 266.

66 Grossman, *On Killing: The Psychological Cost of Learning to Kill in War and Society*, 118.

67 Ibid., 113.

68 Gwynne Dyer, *War: The New Edition* (New York: Random House, 2010), 57.

69 Grossman, *On Killing: The Psychological Cost of Learning to Kill in War and Society*, 108.

70 Dyer, *War: The New Edition*, 87.

71 Grossman, *On Killing: The Psychological Cost of Learning to Kill in War and Society*, 100.

72 Ibid., 101–2.

73 Whetham, 'Remote Killing and Drive-by Wars', 206.

74 Noel Sharkey, 'Saying "No!" To Lethal Autonomous Targeting', *Journal of Military Ethics* 9, no. 4 (2010): 371–2.

75 Singer, *Wired for War: The Robotics Revolution and Conflict in the 21st Century*, 391.

76 Ibid., 392.

77 For instance, see Dale Hanson, 'Predator Strike Video', http://www.youtube.com/watch?v=tIsdwbZSMMw; and DoDvClips, 'Uav Strike on Insurgents', http://www.youtube.com/watch?v=QZ-dNu5uOQc&feature=related. The first clip appears to be raw footage of a Predator strike against enemy forces taking refuge in a typical Middle Eastern stone building. The second is from another unmanned aerial vehicle operated by Coalition forces, which has sighted six heavily armed insurgents on the streets of northern Baghdad and engages them with a hellfire missile. The legitimacy of those targeted is not in question, at least not for our purpose here. In both cases, however, the point being illustrated is that the people being attacked are reduced to little more than 'hot spots' or 'blips' on a black and white screen. It is not clear whether these people, if met by combatants on the ground, would have been spared or simply killed in a similar fashion. What is clear is that no opportunity is afforded to make discriminate and proportional lethal targeting decisions. In both videos, those under attack are located within dense urban areas and are not, at least in the footage made public, actively firing at their attackers.

78 For a much more nuanced account of the different distances, see Tziporah Kasachkoff and John Kleinig, 'Drones, Distance and Death', in *The Rule of Law in an Era of Change: Security, Social Justice and Inclusive Government* (Athens, Greece, 2014).

79 Grossman, *On Killing: The Psychological Cost of Learning to Kill in War and Society*, 161.

80 Peter Watson, *War on the Mind: The Military Uses and Abuses of Psychology* (New York: Basic Books, 1978), 250.

81 Grossman, *On Killing: The Psychological Cost of Learning to Kill in War and Society*, 164.

82 Ibid., see pp. 164–7.

83 Daniel Brunstetter and Megan Braun, 'The Implications of Drones on the Just War Tradition', *Ethics & International Affairs* 25, no. 3 (2011): 339.

84 Ibid., 349.

85 Enemark, *Armed Drones and the Ethics of War: Military Virtue in a Post-Heroic Age*, 46.

86 Ibid.

87 Nolen Gertz, 'Technology and Suffering in War', in *Technology and Security* (University of North Texas: Society for Philosophy and Technology, 2011).

88 Singer, *Wired for War: The Robotics Revolution and Conflict in the 21st Century*.

89 Interestingly, some research has been conducted on using video games to teach just war theory. While there is presumably some merit to this, it seems that any such

effort would really be a matter of overcoming the conditioning that soldiers receive throughout their childhoods. See Tyson Meadors, 'Virtual Jus in Bello: Teaching Just War with Video Games', in *The Ethics of Emerging Military Technologies* (University of San Diego: International Society of Military Ethics, 2011).

90 Lambèr Royakkers and Rinie van Est, 'The Cubicle Warrior: The Marionette of Digitalized Warfare', *Ethics and Information Technology* 12, no. 3 (2010): 291.

91 Grossman, *On Killing: The Psychological Cost of Learning to Kill in War and Society*, 108.

92 Sharkey, 'Saying "No!" To Lethal Autonomous Targeting', 372.

93 See Michael J. Osofsky, Albert Bandura and Philip G. Zimbardo, 'The Role of Moral Disengagement in the Execution Process', *Law and Human Behavior* 29, no. 4 (2005).

94 Sparrow, 'Building a Better Warbot: Ethical Issues in the Design of Unmanned Systems for Military Applications', 175–6.

95 For a complete recount of his experience, see William Manchester, *Goodbye Darkness: A Memoir of the Pacific War* (New York: Dell Publishing Company, 1979), 3–6.

96 Dyer, *War: The New Edition*, 34.

97 Grossman, *On Killing: The Psychological Cost of Learning to Kill in War and Society*, 87.

98 This account of the existence of higher commanders is drawn from David Horner, 'Stress on Higher Commanders in Future Warfare', in *The Human Face of Warfare: Killing, Fear and Chaos in Battle*, ed. Mark Evans and Alan Ryan (St Leonards: Allen & Unwin, 2000), 135.

99 Ibid.

100 Ibid., 139.

101 Ibid., 142–5.

102 Pauline Kaurin, 'Courage Behind a Screen' (2013), 10.

103 Singer, 'The Future of War', 81.

104 Sparrow, 'Building a Better Warbot: Ethical Issues in the Design of Unmanned Systems for Military Applications', 176.

105 Martin and Sasser, *Predator: The Remote-Control Air War over Iraq and Afghanistan: A Pilot's Story*, 209–14.

106 Ibid., 211.

107 For a more comprehensive discussion of this problem as it relates to the professional military ethic, see J. Andrew Ely, 'Drones: A Challenge to the Professional Military Ethic', https://globalecco.org/drones-a-challenge-to-the-professional-military-ethic.

108 Benjamin, *Drone Warfare: Killing by Remote Control*, 97.

109 Chappelle, McDonald, and McMillan, 'Important and Critical Psychological Attributes of USAF MQ-1 Predator and MQ-9 Reaper Pilots According to Subject Matter Experts'.

110 Singer, 'The Future of War', 80.

111 Ibid.

112 Martin and Sasser, *Predator: The Remote-Control Air War over Iraq and Afghanistan: A Pilot's Story*, 306.

113 Ibid.

114 Sparrow, 'Building a Better Warbot: Ethical Issues in the Design of Unmanned Systems for Military Applications', 176.

115 Gertz, 'Technology and Suffering in War'.

116 Singer, *Wired for War: The Robotics Revolution and Conflict in the 21st Century*, 147.

117 Ibid.

118 Singer, 'The Future of War', 80.

119 See Section 2 of the following report for a more detailed breakdown of experience levels amongst unmanned aerial vehicle operators: Chappelle, McDonald, and McMillan, 'Important and Critical Psychological Attributes of USAF MQ-1 Predator and MQ-9 Reaper Pilots According to Subject Matter Experts'.

120 James R. Fitzsimonds and Thomas G. Mahnken, 'Military Officer Attitudes toward UAV Adoption: Exploring Institutional Impediments to Innovation', *Joint Forces Quarterly* 46, no. 3 (2007): 101.

121 Chappelle, McDonald and McMillan, 'Important and Critical Psychological Attributes of USAF MQ-1 Predator and MQ-9 Reaper Pilots According to Subject Matter Experts', 2.

122 Arkin, *Governing Lethal Behavior in Autonomous Robots*, 29–36.

123 Ibid., 31–32.

124 Office of the Surgeon General, 'Mental Health Advisory Team (MHAT) IV Operation Iraqi Freedom 05–07, Final Report' (Washington, DC: United States Department of the Army, 2006).

125 Arkin, *Governing Lethal Behavior in Autonomous Robots*, 36.

126 Mary L. Cummings, 'Automation and Accountability in Decision Support System Interface Design', *Journal of Technology Studies* 32, no. 1 (2006).

127 Ibid.

128 Mary L. Cummings, 'Integrating Ethics in Design through the Value-Sensitive Design Approach', *Science and Engineering Ethics* 12, no. 4 (2006): 710–12.

129 Peter M. Asaro, 'Modeling the Moral User', *IEEE Technology and Society* 28, no. 1 (2009): 22.

130 K. J. W. Craik, 'Theory of the Human Operator in Control Systems I: The Operator as an Engineering System', *British Journal of Psychology* 38, no. 2 (1948); K. J. W. Craik, 'Theory of the Human Operator in Control Systems II: Man as an Element in a Control System', *British Journal of Psychology* 38, no. 3 (1948).

131 Asaro, 'Modeling the Moral User', 23.

132 Ibid.

133 Ibid.

134 Arkin, *Governing Lethal Behavior in Autonomous Robots*.

135 Ibid., 125.

136 Ibid.

137 Ibid.

138 Ibid., 125–6.

139 Wendell Wallach and Colin Allen, *Moral Machines: Teaching Robots Right from Wrong* (Oxford: Oxford University Press, 2009).

140 Andreas Matthias, 'Is the Concept of an Ethical Governor Philosophically Sound?', in *Tilting Perspectives: Technologies on the Stand: Legal and Ethical Questions in Neuroscience and Robotics* (Netherlands: Tilburg University, 2011), 7.

141 Asaro, 'Modeling the Moral User', 21.

142 Lucas, 'Industrial Challenges of Military Robotics', 283.

143 Ibid.

144 Sparrow, 'Robotic Weapons and the Future of War', 122–3.

145 Arkin, *Governing Lethal Behavior in Autonomous Robots*, 177–209.

146 Gordon E. Moore, 'Cramming More Components onto Integrated Circuits', *Electronics* 38, no. 8 (1965).

147 Ray Kurzweil, *The Singularity Is Near: When Humans Transcend Biology* (New York: Viking Penguin, 2005), 138.

148 Joseph Weizenbaum, *Computer Power and Human Reason: From Judgement to Calculation* (New York: W.H. Freeman & Company, 1977), 223.

149 John Searle, 'Minds, Brains, and Programs', *The Behavioral and Brain Sciences* 3, no. 3 (1980).

150 Sullins, 'Robowarfare: Can Robots Be More Ethical Than Humans on the Battlefield?', 274.

151 Bradley Jay Strawser, 'Two Bad Arguments for the Justification of Autonomous Weapons', in *Technology and Security* (University of North Texas: Society for Philosophy and Technology, 2011).

Chapter 7
The Asymmetry Objection

Moving on from discussions of risk in a technological, operational and psychological context, we must now consider the role of unmanned systems and risk at the more strategic interstate level. The game of chess has long been seen as a simulacrum for political and military confrontation at the international level. Whilst clearly a metaphor, it embodies a conception of a very particular type of war and, moreover, a conception that has a great deal of significance for our moral assessment of unmanned systems. When we think of chess, we imagine equally configured forces ready to engage in a perfectly symmetrical contest. Each side has clear and distinguishable uniforms. The battle is regulated by robust rules that stipulate how the conflict is to be commenced, conducted and terminated. As David Rodin argued in his exploration of the ethics of asymmetric conflict, this image reflects a moral assessment of war in two ways. First, it gives us the idea of war as a fair fight between two combatants. Second, because the battle is isolated from all non-combatant elements, it accords with our sense of justice in war, i.e. the risk of harm to those not directly involved in the conflict is non-existent.[1] However, as he points out, there are forms of war that do not embody the sort of symmetry and equality that characterises the contest that is chess. As modern history confirms, war diverges all too often from this chessboard image. This chapter will argue that, when the degree of divergence reaches a critical point, we begin to experience serious difficulties when interpreting and applying just war theory. More specifically, it will be argued that unmanned systems generate a morally problematic 'radical asymmetry' which brings justice and fairness into conflict, or at least into competition, with the aims of drone wars. This chapter begins by clarifying the concept of asymmetric warfare and, in seeking to call into question the legitimacy of radically asymmetric conflicts, it focuses subsequently on the contrasting views of B. J. Strawser and Suzy Killmister. Finally, it will be demonstrated that unmanned warfare challenges the traditional understanding of combatant status and perhaps even some elements of the nature of war.

Clarifying the Concept of Asymmetric Warfare

'Asymmetry' and 'asymmetric warfare' are terms that are used and acknowledged widely throughout military, security and policy communities. US Major General

Perry Smith puts it well in saying that '[asymmetry] is the term of the day'.[2] The problem is that references to asymmetry and associated terms have become so common and casual – to the point that they are virtually omnipresent in scholarly work, government reports and media briefs related to modern military affairs – that there is now a fair deal of confusion and distortion in thinking about asymmetric warfare and this can skew the argument concerning unmanned systems, if not resolved.

While familiar in common parlance, when we begin to apply the terms 'symmetry' and 'asymmetry' to war they take on an additional military meaning such that the definitions and concepts become somewhat less clear. Some argue that asymmetry as a modern military concept did not make its first significant appearance in print until the early to mid 1990s,[3] but detailed references to the same concept can be found at least some 20 years earlier in Andrew Mack's article 'Why Big Nations Lose Small Wars: The Politics of Asymmetric Conflict' in *World Politics*.[4] It was in this article that the term 'asymmetric conflict' was described in detail and through which the concept popularised. As the title implies, Mack was concerned with why large industrial powers failed to achieve victory in conflicts such as those in Aden, Algeria, Cyprus, Indochina, Indonesia, Morocco, Tunisia, Vietnam and others, despite conventional military and technological superiority. To be more precise, he wanted an explanation as to how the militarily powerful could be defeated in armed conflict by the militarily weak. How could the weak win wars? He hypothesised that there must be a range of what he called 'asymmetries' at play. In doing so, Mack acknowledged the work of others who had also written about the role of asymmetries, although in somewhat different terms and with different emphases. For instance, he highlighted that Steven Rosen, E. L. Katzenbach, Johan Galtung and Henry Kissinger have all written about asymmetry in terms of willingness to suffer costs, financial resources, technological resources, goals and strategy.[5] Mack, however, thought that the important asymmetry in the majority of cases was that of public support for political action.[6]

This article went largely ignored for many years, but toward the end of the Cold War period, it had gained renewed interest among concerned academics because of the changing character of war and military conflict. During the 1990s, research building on Mack's insights began to mature. However, it was not until September 11, 2001, that the notion of asymmetry came to the fore, both in the literature and in common discourse. How could a non-state actor successfully carry out an attack that would cripple the United States for weeks and instil long lasting fear in its population? In order to make sense of the events of that day and their aftermath, key political figures evoked the concept of asymmetry. The key user at the time was Secretary of Defense Donald Rumsfeld. After noting that the United States was 'going to have to fashion a new vocabulary' to describe this 'new' kind of warfare, he told reporters that he had long been talking of 'asymmetrical threats'. When a savvy reporter pressed the Secretary for further information about what these asymmetrical threats actually were, he could not provide a definition.[7] However, a working definition of 'asymmetric warfare' published by the US Joint Chiefs of Staff states that:

> Asymmetric approaches are attempts to circumvent or undermine US strengths while exploiting US weaknesses using methods that differ significantly from the United States' expected method of operations.[8]

From this definition, it is clear that asymmetric warfare is perceived as a strategic threat to the United States and one that may, in theory, be present in any number of different guises and conflict areas. Rumsfeld was able to clarify what he thought to be asymmetric tactics by offering some examples: 'terrorism and ballistic missiles, cruise missiles, [and] cyberattacks'.[9] Others have provided more comprehensive lists. For example, Kenneth McKenzie, again approaching the matter from a US strategic context, identifies six categories of asymmetric tactics. These include the use of or deployment of: 1) chemical weapons; 2) biological weapons; 3) nuclear weapons (known collectively in the post-Iraq era as 'weapons of mass destruction'); 4) information war (such as attacks on key financial, infrastructure or defence systems); 5) terrorism (which is notable because of its focus on non-combatants); and 6) other operational concepts that may involve guerrilla tactics, the involvement of non-state actors, the commingling of military forces with civilian communities in an effort to complicate weapons use, and the use of primitive weapons in unusual and surprising ways.[10] These US-centric descriptions, when taken in their historical context, reflect the more general fact that the term 'asymmetric warfare' has become synonymous with unconventional attacks which leverage the vulnerabilities of the strong (that are either overlooked or tolerated) in order to avoid conventional military engagements, particularly with Western powers. Rodin confirms this, writing that 'asymmetric tactics are typically the tactics of weakness, not tactics of choice; they are adopted by those who do not have the military capability to engage their enemy on roughly equal terms in conventional war'.[11]

It is wise to take issue with this United States-centric conceptualisation of asymmetric warfare on the basis that it is inappropriately constricting, as will soon be demonstrated. It should first be pointed out that all wars are asymmetric in the sense that perfect equality is highly unlikely to occur in the modern military world and, in any case, it would be practically impossible to verify whether there is perfect equality.[12] This is not to deny that the definition given earlier is useful in describing the current state of military affairs, i.e. in which there are many conventionally weak but potentially dangerous adversaries prepared to confront world superpowers like the United States with such tactics. However, there remains a need for a sharper and more accurate understanding of asymmetry in warfare. As a bare minimum, a more acceptable conception must establish asymmetry in warfare as a category independent of the perceived power of the actors involved. Let us think back to Mack. While he framed his discussion in terms of how the weak could prevail against the strong, all the asymmetries he identified were simply disparities of some sort between opposing actors in a conflict. The same can be said of those identified by McKenzie. Indeed, effective chemical, biological, radiological and nuclear weapons are only in the domain of the strong, which leads to the thought that asymmetric war is not something that can only be waged by the weak. While weaker actors may have greater reason for waging asymmetric warfare, asymmetric tactics are often a tool of the strong. Furthermore, as Enemark argues, it needs to be recognised that 'asymmetry' can refer to any imbalance in strength that each side in a conflict can bring to bear against the other.[13] Another way to conceive of the matter is to assume that there are positive asymmetries, which involve utilising differences to gain an advantage, as well as negative asymmetries, which might be used by an opponent to gain an advantage.[14]

That said, it must finally be stressed that there are many asymmetries at play in the drone use that has now spread throughout many countries over a period of more than a decade. The most obvious asymmetry, of course, is technological in nature. This is part of a more general asymmetric relationship generated by drone operations, which should, in Mack's terms, be classified as a function of the asymmetry in 'resource power'.[15] As will be shown in the next section, an overwhelming asymmetry in technological resource power may also cast doubt over the moral legitimacy of military action. It must also be said that, while technological strength may have guaranteed decisive military victory in earlier years, weaker opponents have since realised that they are not compelled to fight their enemies on the dictated terms and can utilise any of those asymmetries mentioned above. In the remaining sections, it will be shown how these diverse asymmetries further complicate drone warfare.

Radical Technological Asymmetry

As suggested earlier, technological asymmetry is nothing new, at least in the outdated weak-versus-strong sense. Despite this, we can highlight a few conflicts that illustrate humanity's progression along the scale of technological asymmetry. The Battle of Agincourt is often cited as a striking historical example of the sort of effect superior military technology can have on the battlefield. It was at Agincourt that the English achieved victory over the numerically superior French army. It was thought that the English were outnumbered by almost six to one, with the French having troop strength of approximately

30,000 men.[16] Yet what is striking is that due to the longbow, which allowed its users to launch a barrage of powerful arrows at a greater than normal distance, the English suffered only a few hundred war dead, while the French lost many thousands.[17] William Shakespeare later depicted this conflict as ruthless and unchivalrous, a depiction which reflects some of the concerns of this chapter. According to Shakespeare, ruthless and ungallant conflict was essentially one of the paradoxes for the king; something he must wrestle with when protecting his empire.[18] Another engagement representative of technological asymmetry was that which took place between well-armed colonial forces and an army of traditional African warriors at Omdurman, on the upper Nile, at the end of the nineteenth century.[19] Under the command of General Horatio Kitchener, the colonial forces fought a vastly larger armed force and managed to achieve a decisive victory, which was critical to the British conquering Sudan. The Egyptian forces under Kitchener's command carried breach-loading rifles and the British carried repeating rifles (which allowed for faster firing), maxims (machine guns) and field cannon. Their opponents, on the other hand, had only spears, swords and muskets.[20] Unsurprisingly, the Anglo-Egyptian casualties were few and far between, while their opponents lost approximately 10,000 troops.[21] Showing his concern about asymmetric war, Winston Churchill, who fought in this conflict, later wrote about his disillusionment with modern technology and how dishonourable the Omdurman 'slaughter' was.[22]

The Gulf War was another key point in the progression toward radical technological asymmetry in the more modern sense of the term. Executed by a United Nations coalition force, led by the United States and financed mostly by Saudi Arabia, the Gulf War was waged against Iraq primarily in response to its invasion and annexation of Kuwait. While the Iraq of the 1990s was far from a defenceless state, its weaponry was simply not as advanced as that of the Coalition.[23] The Gulf War saw the introduction of advanced networked technologies including fighter jets, surveillance aircraft and so on. It was widely portrayed as a 'telepresent' war conducted by 'armchair generals' via satellite.[24] There was little contest in this war, with the Coalition achieving a decisive victory with very few losses. The Gulf War essentially served as a template for conducting modern asymmetric warfare, and the Kosovo War was its corollary. This war, aimed at bringing down the Milosevic regime and protecting Kosovo from Serbian aggression, began with several United States ships and a British submarine (operating under the North Atlantic Treaty Organization, NATO) firing cruise missiles. Following this, allied aircraft launched a campaign of precision-guided munitions attacks, which were aided by the use of drones.[25] Given the technological asymmetry that existed, NATO's aims were met with no military casualties on the allied forces' side.[26] Accordingly, some theorists have labelled both Kosovo and Iraq as being 'no risk' wars. In both cases, this is an obvious misnomer. No war, at least no war that we can conceive of today, can be totally risk free. War will always pose *some* harm to both combatants and non-combatants. What these theorists mean to convey is that the states which wage this sort of technologically asymmetric warfare are taking significantly less risk, and that such warfare has progressed significantly from the level playing field of the chessboard, which was discussed earlier.

If technological asymmetry is already a common feature of modern warfare as suggested, one might then wonder whether the technological asymmetry generated by unmanned systems really represents a 'radical' departure from previous levels of asymmetry. Is firing a missile from a drone more ethically problematic than dropping a 500 lb bomb at 20,000 ft or pressing a button that launches a cruise missile? The answer is straightforwardly: 'yes'. What makes the former morally repugnant is the ability to target a specific individual, at great distance, with the knowledge that you are invulnerable to retaliation. Unmanned systems make it possible to remove the human actor from the area of conflict – all the while allowing the operators of these systems to target people and infrastructure almost as if they were present in the field themselves – taking us as close to the notion of a 'risk free' warfare as we are likely to get at any time in the foreseeable future. This is unlike the conflicts with Kosovo and Iraq, in which pilots remained in the air, thus somewhat putting themselves at risk. Objectors to this sort of risk free war say that an issue arises when unmanned systems are used against another force, which does not possess such technology, and when the level of life-threatening risk incurred between the warring parties becomes so imbalanced that we cross a symmetry threshold which makes the fight (and thus the use of the weapon) intrinsically unjust. To illustrate what people find problematic and to test our moral intuitions, let us consider the following thought experiment.

State X, holding what it thinks is a just cause, declares war on State Y. State X possesses an arsenal of unmanned systems. Rather than sending in many traditional human warfighters, State X sends in its unmanned weaponry, starting with reconnaissance drones (ranging from wasp sized micro-unmanned aerial vehicles to more traditional high altitude vehicles) to surveil the area and provide intelligence. Once this information is transferred wirelessly across the military network and analysed by super-computers, State X sends in various drones armed with technologically sophisticated munitions to kill the combatants and destroy the infrastructure of State Y. State Y, for whatever reason, does not have an arsenal of unmanned weaponry and has no enemy combatants within range of its conventional weaponry, since its enemy has a purely technological presence. So, at the moment at which State X commences their remote attack, State Y essentially is doomed to fail in defending itself.[27]

It is this kind of scenario – where one side can inflict damage on the other with virtual impunity – which ethicists find morally troubling. The issue is how such unmanned conflict could possibly be considered ethical. There are two grounds on which the legitimacy of asymmetrical unmanned warfare might be questioned and these will be considered separately. The first has to do with the notion of fairness and equality in warfare, and how it might be undermined by radical technological asymmetry. The second, which is equally important, has to do with what will be called 'evoked potential': that is, the spontaneous, potentially dangerous and morally questionable alternative asymmetric response/s possibly evoked by the sort of technological asymmetry which is of concern to us. This potential can apply to both state and non-state actors. It is relevant to note here that there are all sorts of unresolved debates concerning the difference between these two sorts of actors, and whether non-state actors can act in accordance with just war principles in the same fashion as state actors. Here though, the discussion is restricted primarily to illustrating the moral problem with radical technological asymmetry and makes no claims about this sub-debate.

Fairness in the Absence of Mutual Risk

Strawser presents the most comprehensive account of the issues of fairness as they relate to unmanned systems. However, given that he discusses these issues as a rebuttal to the case *against* unmanned systems, we first need to briefly recap his case *for* unmanned systems, which was outlined in Chapter 3. Strawser claims that if it is possible for a military pursuing a just war sanctioned action to use unmanned systems in lieu of the manned equivalent, without incurring a loss of capability (by which he means *just war* fighting capability), then they have an ethical obligation to do so. This contention stems from what Strawser labels the 'principle of unnecessary risk', which holds that: in trying to accomplish some objectively good goal, one must, *ceteris paribus*, choose means that do not violate the demands of justice, make the world worse or entail more risk than necessary to achieve the good goal in question.[28] This principle has a fair deal of *prima facie* appeal. When we apply it to warfare with unmanned systems, it seems rather uncontroversial to suggest that any technology that *exclusively* minimises harm to warfighters – that is, while imposing no additional 'costs' of another kind – would be morally better. Notwithstanding its *prima facie* appeal, the problem with this principle's application is that, while unmanned systems may indeed minimise immediate harm or the risk of it, there may be other unforeseen consequences that countervail the ethical obligation for their use. The introduction of radical asymmetry to unmanned warfare is one such countervailing factor.

Strawser responds to the fairness problem using two arguments, the first of which is perhaps less substantial than the second.[29] His first response is to say that, if the issue concerns whether justified combat should be a 'fair fight' to some extent, this presents no problem for the use of unmanned systems, in principle, because military engagements have been anything but fair for quite some time. He gives the example of a modern day F-22 pilot killing a tribal warrior wielding a rocket-propelled grenade.[30] However, this example fails to support Strawser's response. While Omdurman, Iraq and Kosovo all confirm that unfairness is not new, it can be argued along the lines presented in the previous section that radically asymmetric unmanned conflicts introduce a new, or at least differently disturbing, level of unfairness. In the above example involving the F-22, there is a pilot in the air. Therefore the tribal warrior still has a human to target, regardless of how futile his/her efforts to kill that pilot may be. By contrast, the introduction of unmanned platforms removes most – if not all – warfighters from the field, and this allows them to overwhelm the enemy at no comparable risk to themselves, using purely technological means.

Strawser's second, and main, line of reply to the fairness objection is that, even if unmanned systems *can* be said to have introduced a new and disturbing level of asymmetry, this still does not present a significant problem for their use.[31] His reasoning appears to be the following: if one combatant is ethically justified in their effort and the other combatant is not, then it is good and just that the justified combatant has the asymmetrical advantage associated with the use of unmanned systems. In his view, this is because combatants fighting for an objectively unjust cause have no moral right to engage in violent action, or even defensive action, against combatants fighting for an objectively just cause. There is a moral difference between the two, he would say. Here, Strawser is invoking recent work by Jeff McMahan,[32] which presents a fairly novel reinterpretation of classical just war theory. While it is not clear whether McMahan's work ultimately has a significant impact on the central issue, it is nonetheless worthwhile working through his response thoroughly. Doing so will provide context for our overall concern with asymmetry, which is a concern regarding the justice of resorting to war in such scenarios. For the ensuing discussion, it will be helpful to briefly recount the traditional theory of just war and the challenges that McMahan's revision raise for it.

Recall that the traditional theory of just war embodies two main sets of principles that provide a rigorous moral framework. The first concerns the justice of the resort to war (*jus ad bellum*) and the second concerns just and fair conduct in war (*jus in bello*). Under the *jus ad bellum* principles, a state intending to wage war must possess just cause, right intention and proper authority. The war must also be proportional, a likely success and a last resort. Under *jus in bello* principles, activities in war must be discriminate and again, proportional.[33] Under the traditional just war theory, these two central pillars are considered logically independent from one another. Most importantly for our discussion, this means that it is permissible for a combatant fighting for an objectively unjust cause (an unjust combatant), to fight against a combatant fighting for an objectively just cause (a just combatant), as long as they do not violate the principles of *jus in bello*. In other words, just and unjust combatants are taken to share the same moral status in war. This is commonly referred to as 'the doctrine of the moral equality of combatants'.[34] It is essentially a doctrine of battlefield equality. The reason combatants are not held responsible for the decision *to go to war* under this doctrine is because they may, among other things, have been under duress or lacked the requisite information to determine whether their cause was genuinely just, so they are instead judged exclusively on how they fight in the ensuing conflict.

Importantly, Strawser, following McMahan, would reject this notion of equality between combatants. McMahan claims that it seems counterintuitive to say that those pursuing an objectively unjust aim are morally on par with those pursuing an objectively just aim. It is not as though McMahan thinks that we should abandon just war theory altogether. Rather, he argues that the principles of *jus in bello* should not be considered to be logically independent of those of *jus ad bellum*. Remember that the *jus in bello* convention requires that fighting in war be both discriminate and proportional. McMahan believes that it is virtually impossible to fight in a war, while lacking a just cause, without violating one of these principles.[35] Put most simply, he says that unjust combatants breach the discrimination principle because just combatants are not legitimate targets, since they are innocent in a special sense. Just as a police officer retains their innocence when apprehending a lawbreaker, a just combatant retains her/his innocence in fighting an unjust combatant. Unjust combatants also breach the proportionality principle, because without a just cause, there is no objective good to outweigh any harm done. So, for McMahan, this is why the *jus in bello* principles must be considered in relation to the *jus ad bellum* principles, and also why combatants are not considered to be moral equals in war. Let us label this the 'non-equality thesis'.

Strawser argues from the non-equality thesis that it is good that the just combatant has the asymmetric advantage associated with the use of unmanned systems and is better protected. For Strawser, overwhelming the enemy with a barrage of drones and condemning them to what is going to be certain defeat is not a morally objectionable act. This is because the just combatant is morally justified in taking the life of the unjust combatant, while the unjust combatant is not justified in taking the life of the just combatant, even if the unjust combatant appears to fight in accordance with *jus in bello*. Therefore, according to Strawser, if a particular military force fighting for a justified cause has a better, more advanced and more effective weapons systems than that of their unjustified adversary, they should not refrain from using it because it is seen as 'unfair' or 'unchivalrous'. They are, by the above reasoning, justified in getting that weapons system into the field as quickly as they can to aid in force preservation measures and to improve the likelihood of winning. Thus, responding to the critic's question: 'how can this war be just?', Strawser would say that unjust combatants, who are incapable of fulfilling the requirements of *jus in bello*, are owed no special consideration when it comes to using unmanned systems. This is because they are contributing to a moral wrong, whether or not they are consciously aware of it.

Like Walzer,[36] many will not be convinced by the non-equality thesis, which underpins Strawser's denial that the asymmetry poses a problem for drone employment. The argument proposed is that, in line with traditional thought on the topic, *jus ad bellum* and *jus in bello* should be thought of as logically independent and that we are correct in making the *prima facie* presumption that there exists moral equality between combatants. As Walzer so eloquently puts it:

> What Jeff McMahan means to provide in this essay is a careful and precise account of individual responsibility in time of war. What he actually provides, I think, is a careful and precise account of what individual responsibility in war would be like if war were a peacetime activity.[37]

These comments highlight the following important point: there is something about war that makes the moral standards that apply to it different from those in normal civilian life. That is, there is something about war, which permits soldiers to do things that are normally considered immoral. There are numerous reasons as to why this might be so. Many theorists think it fit to talk in terms of the forfeiture of rights, i.e., soldiers waiving rights they would normally hold in peacetime. However, there are a number of deep problems connected to this approach, particularly concerning the inalienability of certain rights, namely that to life. Another reason, which follows from what was said in chapters 3 and 4, is that there is something in the nature of the imperfect relationship between individuals and states which allows for those on both sides to fight justly regardless of the objective justness of their cause. But, for the moment, let us suppose that this is wrong so that unjust combatants do, in fact, act wrongly when targeting just combatants, and therefore the just combatants are entitled to defend themselves against the overall wrong being perpetrated by the unjust combatants. It is not entirely clear how relevant this actually is to the problem. In other words, it may be that these questions of fairness and equality – at the *jus in bello* level at which they are raised by Strawser – are insufficient to explain or deal with what is fundamentally problematic about the asymmetry in question.

A deeper problem for those advancing the fairness objection consists in pointing out that using unmanned systems in place of the manned equivalent removes an important element of the justice of resorting to war, an element that seems important regardless of whether the non-equality thesis succeeds. More specifically, when the technological imbalance reaches a certain level, it may actually override any justification for war. To both clarify the fairness objection and draw out the issues in greater detail, it will be helpful to look briefly at Paul Kahn's 'paradox of riskless warfare'. This paradox is said to arise when the 'pursuit of asymmetry undermines reciprocity [in terms of risk]'.[38] Kahn says that any fighting between combatants needs to be qualified in terms of the mutual imposition of risk. [39] Kahn's paradox occurs at the level of individual combatants or groups of combatants and is underpinned by a belief in something akin to the moral equality of combatants, which Strawser refutes. However, for the moment the reader can put this matter aside, because the aim in discussing Khan's work is simply to elicit some thought on the general role of risk. Kahn essentially says that, without the reciprocal imposition of risk, the moral basis for injuring or killing others is called into question. It is not that he advocates a concept of war of the chessboard variety, which has equally configured forces. Rather, what he is saying is that the right to engage in potentially lethal conduct only holds where there is some *degree* of mutual risk involved. He seems right in using the notion of risk, and thus threat, as a condition. Sparrow discusses this idea and deploys it at the level of individual responsibility. He notes that in wartime, wounded soldiers are generally considered immune from attack, as they no longer pose an immediate or near-term threat. Similarly, those who raise the white flag and surrender are also considered immune from attack, because they likewise pose no threat.[40] In both cases, threat (or lack thereof) is at the core of their immunity.

A full account of Kahn's ideas cannot be provided here, but it is evident that the degree of threat plays an important role in establishing and maintaining any justification to cause harm or kill. This means is that, if one side's armed forces are incapable, or rendered incapable, of posing an adequate threat to the other side, the more powerful side's moral justification for targeting the weaker side's armed forces is void.[41] Yet, as mentioned earlier, Kahn's argument is grounded at the individual level. He neglects the role of risk at a higher level, but it is exactly this sort of risk that we need to consider in order to refute Strawser's argument. As noted by Sparrow as well, the kind of asymmetry which is relevant here is that which exists at the macro level, namely at the level of armed forces considered as a whole.[42] This is especially true for our discussion of the use of unmanned systems. The examples given earlier, of the soldiers who are wounded or have surrendered, perhaps convey the idea that the sort of threat about which we are talking is located at the micro level, that is, between individual combatants. This is the level at which Strawser's discussion takes place. However, with reference to the earlier scenario, the threat that is relevant here is that between State X and State Y considered as a whole, not that between the individual combatants of State X and State Y. This is an important point, because if there is an inadequate level of risk between States X and Y considered as a whole, any reasons that the individual combatants of these states have for taking on any risk will be far less compelling.

In just war terms, the issue is the following: when the level of asymmetry in war reaches a certain level, a state may be in violation of *jus ad bellum*, the principles of which remain important despite the challenge posed by Strawser's McMahanian argument. Why is this so? Up until this point, we have been talking about the 'deep morality of war'. This is distinct from Walzer's conventional morality of war in that it revolves around the idealistic notion of a system based on objective justice and individual liability. It is this idealistic notion that is responsible for the non-equality thesis and much of Strawser's argument concerning unmanned systems. However, McMahan concedes that there may be laws of war (which embody the traditional just war principles) to which we should adhere for prudential reasons. These principles will, if consistently and accurately observed, tend to limit the destructiveness of war. This is the reason why the *jus ad bellum* principles remain relevant. Yet, McMahan would say that there might be cases in which a given act of violating these *jus ad bellum* principles, when viewed in isolation, might be permitted by the deep morality of war. However, if this act of violation were to lead to other violations, this would make the said act imprudent, because it would fail to limit the destructiveness of war.[43] This two-tiered approach is problematic. As argued in chapters 3 and 4, our just war principles already provide a transitional morality, which will hopefully lead us toward a better state of peace. In granting that we have prudential reasons for adhering to the laws of war, McMahan seems to acknowledge the need for a transitional morality. This seems to mean that his deep morality of war serves little purpose, other than to remind us of the end goal, i.e. a better state of peace.[44] His deep morality of war cannot be allowed to constantly trump the transitional morality of war. A full discussion of the reason why we ought to be reluctant to sharply contract prudence with this more idealistic deep morality is beyond the scope of this book. The point made here, which is rather uncontroversial, consists in suggesting that, even alongside a deep morality of war and the non-equality thesis – and in any warfare, even against the unjust – the just side must adhere to *jus ad bellum* principles. They cannot do whatever they like by appealing to whatever they consider to be the objective justness of their cause.

Having argued that the *jus ad bellum* principles remain important even when fighting an objectively unjust opponent, there are two principles that need to be given particular attention when considering waging a war with unmanned systems. Both principles are grounded in consequentialist considerations (although one can equally well think of them in deontological terms). The first is the principle of last resort. It is generally recognised that, once war is unleashed, it can often get out of control and take on its own destructive power, with devastating consequences. Therefore, war should only ever be fought when necessity demands it; that is, when there is no other alternative. However, waging war without any risk to one's own troops clearly calls into question one's adherence to this principle. Where there are such high levels of asymmetry created by technology, as in our scenario, war surely cannot be considered a last resort. In most cases, State X would presumably have other less lethal options available. For example, State X could make clear and obvious to its opponent the fact that it possesses significantly superior technology, perhaps by putting on a non-lethal display of its technological capabilities. This may result in both States reaching some sort of peaceful political negotiation and settlement. Second, there is the principle of proportionality. In the *jus ad bellum* sense, this principle asks us to look at the harm our proposed military action is likely to cause and weigh it against the harm that we hope to avoid. But, where the technological imbalance is so radical – that is, where those on the receiving side of an attack are virtually unable to retaliate – it seems that the harm that the technologically superior state hopes to thwart will in many cases be so insignificant that it would present dire problems for the proportionality calculus. In other words, unmanned war of the sort described in the earlier scenario is rendered unjust. Of course, the deployment of unmanned force can be more limited in nature and need not cross the symmetry threshold, but the escalation of war poses a constant risk.

Evoked Potential

For the reasons described above, war does not necessarily need to be a 'fair' or perfectly symmetrical fight, as represented by the game of chess. But war arguably needs to be *a fight* of some description if it is to have any hope of fulfilling the *jus ad bellum* principles of last resort and proportionality.[45] Regrettably, it is not immediately clear whether the recent drone wars meet this description. In fact, in many ways, their employment seems to facilitate the waging of politically motivated violence that, as it turns out, more closely resembles extermination. This section aims to outline a potential counter argument to the asymmetry objection or, more particularly, to the principle relating to proportionality. The other principle, which pertains to matters of last resort, will remain untouched. In pre-empting the asymmetry objection and clarifying what exactly is at issue, it will be shown that we are presented with another potential problem for radically asymmetric unmanned warfare and thus have a further reason to question its legitimacy.

As we already know, at the *jus ad bellum* level, the principle of proportionality requires that we weigh the harm that the proposed military action is likely to cause with the harm that we hope to avoid. Highlighted in the previous section was the issue that the harm faced was not great enough to justify the attack in question. However, one might object that the only reason the harm was not seen to be so significant was because we failed to look beyond the initial pre-provocation harm and failed to accurately project the longer-term consequences. It might be said that a state contemplating waging war must think more clearly, when projecting the outcomes of a possible conflict, about how its actions will impact and/or provoke the enemy – including how its actions and decisions will influence the enemy's will and response. In other words, the stronger state ought to factor in evoked potential: the spontaneous, possibly dangerous and sometimes morally questionable responses caused by radical technological asymmetry. The problem is, of course, that the extent to which we are required to project consequences under the proportionality principle is not obvious. After all, it is remarkably difficult to predict what the enemy's decisions and responses will be prior to initiating a conflict and traditionally, they have not been given much weight.[46] However, since evoked potential is indirectly linked to radical technological asymmetry, and given that states have an obvious interest in not being subjected to asymmetric tactics, we will, for the sake of argument, entertain the idea that a state should foresee such responses and include them in its calculation of the proportionality of its proposed actions. If nothing else, in dealing with the objection, we should actually demonstrate – presumably to the dismay of those who would object to the earlier treatment of the fairness problem – that evoked potential might tip the proportionality scale too far in the other direction. That is, if we must consider evoked potential, a radically asymmetric attack may involve so much potential harm that the war will be considered unjust.

In order to explore the likely responses from a state under radically asymmetric unmanned attack, let us revisit and develop the scenario sketched earlier. State X, a democratic state having what it takes to be a just cause, wages war against an undeveloped State Y. It seems that State Y, which disputes the claims underlying State X's just cause, has a right to try and defend itself. State X wages war with unmanned systems, but because State Y lacks its own arsenal of unmanned weaponry, it cannot respond in kind. Furthermore, because the war is of the unmanned variety, State Y cannot respond in what has traditionally been thought to be the 'normal way' – i.e. by targeting the troops on the ground that are engaged in the 'business of war'.[47] In exercising its right to self-defence, State Y wants to act in an ethically sanctioned manner, because it is aware that it derives its power from its people, who have a vested interest in acting morally under the social contract that assures their continued wellbeing. But State Y is limited in its pursuit of just war and general ethical aims by its under-developed technology, and it must therefore seek alternative ways of responding to the unmanned attack.[48] These alternatives are what we must work through to form some idea of what State Y's response and its probable effects are likely to be, and whether, in light of this forward-looking projection, a radically asymmetric attack remains the most effective and ethical option for State X to achieve its desired goal/s.

Killmister argues that, once we have ruled out the extraordinarily unlikely option of unconditional surrender, there is only a limited number of options open to the weaker state in a situation where there exists a radical imbalance in technological resource power.[49] In the aforementioned example, the vast technological difference means that State X can quite easily locate and attack State Y's combatants just because they are likely to be wearing uniforms as well as bearing arms that have easily identifiable characteristics or 'signatures'. The same task is made rather more difficult for State Y because of their enemy's purely technological presence. This gives rise to an ethically concerning potentiality which Bob Simpson and Rob Sparrow have labelled the 'guerrilla problem'.[50] In order to avoid being targeted and possibly killed by their opponents who have better surveillance and attack capacities, the combatants of State Y may embrace a common asymmetric tactic and decide that, rather than carry arms openly, they will try to conceal themselves and their weapons amongst the vulnerable civilian population. This tactic has the effect of making it much more difficult and potentially dangerous for State X to abide by the just war principles of discrimination and proportionality. For conventionally weak states, the aim is often to provoke the enemy and then commingle with the civilian population in the hope that the enemy will be driven to continue with attacks causing a large number of civilian casualties and other collateral damage. This will, in turn, undermine both local and international political support for the stronger adversary's cause, as will be seen in the next chapter.

At this juncture, it is important to note that the conventionally weak State Y is not alone in utilising this asymmetric tactic and in shifting the burden of risk onto its civilian population. Drone pilots are regularly commingled with civilians. The most well-known US drone command centre operates from the CIA's headquarters in Langley, Virginia, whilst Nevada's Creech and Nellis Air Force bases near Las Vegas house similar military operations. Domestic bases also operate in California, Arizona, New Mexico, North and South Dakota, Missouri, Ohio, New York and perhaps elsewhere.[51] In early 2013, Fox News reported that new jobs and drone technology were coming to the former Willow Grove Naval Air Station in Horsham, Pennsylvania.[52] But do drone pilots count as combatants and therefore as appropriate targets for their enemies? This is an important question because, while a good deal has been said about the psychological effects of commingling and protecting those who may be vulnerable to drone strikes, little has been said about protecting the civilians amongst the operators of unmanned platforms. A drone operator who actively tracks and kills enemies is a possible target, because s/he is a participant in combat. However, for a large number of drone operators, their civilian and military worlds are intertwined. Problems about their status arise when an operator finishes her/his shift and goes home. We need to think carefully about whether drone operators are targetable when they are eating, sleeping and picking up the children from school. Conventional combatants do not acquire immunity when they eat or sleep and cannot simply 'turn off' their combatant status. But, as drones proliferate around the world and wars become an increasingly part-time endeavour for those engaged in fighting them, we may be forced to reconsider the characteristics and nature of combatant status.

Whether or not we consider State X's drone pilots to be combatants, any story of the sort of technological asymmetry that accompanies the widespread deployment of unmanned systems and the shielding of military assets will necessarily include those in the civilian realm being exposed to what might be perceived to be an unfair level of risk. This is a recurring theme in Andrew Croome's *Midnight Empire*,[53] an espionage thriller that deftly explores drone warfare. In this novel of modern warfare, an Australian drone programmer relocates to the United States to work with a drone team operating at Creech Air Force base. Soon after he arrives, however, American drone pilots start to die in the usually peaceful residential suburbs of Las Vegas. As it turns out, they are being targeted by their targets. The point highlighted by the novel is that, far from the promise of delivering a safer and more effective remote warfare, unmanned systems may actually bring the battle home by putting those who wage it closer to us than ever before. Over the long term, this may help foster sympathy between the citizens of warring states, but, in the near future, it is likely to inspire hatred. Additionally, although the killing of drone pilots may not be considered a violation of present-day just war theory, radical asymmetry may come cause anger and actions that more closely resemble classical domestic terrorism. In fact, the evidence suggests that drone strikes only strengthen the terrorist cause, making the al-Qaeda brand and radical Islam more attractive to vulnerable and disaffected American Muslims. For instance, there were two young Boston Marathon bombers who cited the drone wars in Iraq and Afghanistan as a motivating factor for their terrorist action.[54] Cases like this suggest that we need to take seriously the idea, which will be put forward in the next chapter, that drones (as currently used) create more terrorists than the number of enemy combatants they kill.

Soldiers and statesmen must, therefore, add a corollary to the defence theorem: if one side uses a large fleet of unmanned systems against which the enemy has no real defence, the victims may reciprocate by utilising tactics that render the technologically superior state comparably defenceless. It is not immediately clear whether these tactics will, in all cases, be contrary to the demands of just war theory or unfair on civilians. As mentioned earlier and regarding the first point, there is much debate concerning whether terrorist action can meet just war requirements.[55] As for the second point, it needs to be admitted that citizens of technologically advanced democratic countries are on the one hand the ultimate source of the antagonism towards casualty, but on the other hand they are responsible for electing those who authorise war.

Citizens may thus be forced to accept that the aforementioned evoked potential is simply a concerning, but necessary feature, of the only sort of warfare they are prepared to sanction. That said, it is the case that, in radically asymmetric conflict, violations of *jus in bello* are much more likely to occur on either side of the conflict. In relation to our scenario, this may tip the proportionality scale against the waging of war in the first instance. That is, if State Y's responses can be foreseen, it may again undermine the *jus ad bellum* justification for State X's actions. If there is cause to suggest that provoking State Y might cause it to appeal to the abovementioned morally questionable options, State X must think more carefully about what it stands to gain from going to war.

Some might object to the level of foresight required by this account. Let us summarise the argument for simplicity's sake: State Y is doing this bad thing (prosecuting a war with what its opponent sees as unjust aims) and then, because State X utilises drones as their means of waging war, State Y to does this further bad thing (in this case, harming people near to the place/s from where the drones are operated). It might be said that this is not an argument against State X. The critic might also say that this sort of moral reasoning is pretty strained as an argument against the use of unmanned systems as a just weapon of war. It has been said that a very high degree of certainty would be required about the fact that using drones (an action which has other normative advantages) would necessarily cause State Y to do this further bad thing.[56] Furthermore, it has been argued that, even if it can be assured that State Y will do this further bad thing, the moral blame still falls on Y for doing this bad thing, not on X for carrying out a putatively just action via putatively just means. In response to the first section of this argument, it is only necessary to point out that the degree of epistemic certainty involved is already high. History serves as a potent reminder of the fact that technological asymmetry does not preclude enemies with inferior technology from creating and sustaining an impact powerful enough to combat a superpower. In fact, technological asymmetry probably invites such a response from technologically inferior foes, since there are no other options for them.

This does not amount to reassigning blame to State X and absolving State Y of moral responsibility for its actions, nor does it open the door for any state to do bad things in the hope that they might get their enemy to cease efforts against them. Rather, this amounts to acknowledging that the harm generated by such wars comes about for predictable and identifiable reasons. This constitutes the first step toward reaching more amicable resolutions to conflicts in the future. Note that this need not be a big problem for Strawser's main claim *per se*. Recall what his claim was: if it is possible for a military pursuing a *just action* to use unmanned systems in lieu of the manned equivalent, without incurring a loss of *just war fighting capability*, then they ought to do so. Strawser could then simply accept that, *if* unmanned systems do make military action unjust, they should *not* be used. This may be true in some cases of radically asymmetric conflict. But the answer is not necessarily crystal clear in all cases. These systems are not yet beyond control or restriction, so we are left in quite a tricky position as to how we might seek to respond to the problem of radically imbalanced levels of mutual risk, regardless of whether the levels of risk are inadequate or too substantial. We do not want to rule out military operations in every instance. This idea is untenable under the military-state contract and would be rather unwise, because to lay down arms could create an even worse asymmetry problem. The relevant question is now how unmanned systems should be used when the levels of risk are significantly imbalanced. It is a question to which we need to devote more of our attention given the ever-growing technological imbalance between states.

One possible response might be to suggest that the militarily dominant state (possessing unmanned systems) allows the weaker state a degree of latitude in their application of the just war principles. However, allowing an enemy to choose an action, which is not as good/ethical as another one, seems counterintuitive and may lead to serious moral transgressions. For instance, it would be wrong for a strong state to allow its weaker enemy to a carryout disproportionate attack because of technological asymmetry. Therefore, we can say with confidence that giving the weaker state a 'wild card', so to speak, is not the answer. Another much more respectable and less problematic approach, advocated by David Rodin, is for the militarily dominant state to impose stronger requirements on its own application/s or consideration of the just war principles.[57] This recommends that the stronger state has to meet higher standards of epistemic certainty when waging war. It seems a sensible ethical option. It does not encourage the stronger state to lower protective measures, nor to 'go easy' on the enemy, nor to lay down its arms, which is good because there is no virtue in taking risk purely for the sake of risk. What, then, might a war that abides by these stronger norms look like? Kahn suggests that, in cases of such high asymmetry, the solution is for the application of any military force to be very restrained and for it to be more like a form of international policing.[58] Such action is based on evidence proving guilt beyond reasonable doubt. De-escalation strategies are also to be preferred; very little collateral damage is tolerable and the death of innocents is strictly prohibited. The effect that policing actions have would amount to giving noncombatants a much higher standing than that of drone operators, and would imply that the number of strikes is to be limited. However, while fewer wars and more international policing-like activity may sound like an admirable goal, some would argue that, as a response to asymmetry, this is too restrictive and we must continue to find the right balance. It may be that if cases of asymmetrical conflict should be governed by standards that are more restrictive than those guiding interstate war, but more permissive than those of domestic law enforcement, we should explore the category of *jus ad vim* (typically concerning justice of force short of war) to assess the ethical use of drones. However, this is ultimately a matter that must be left for discussion by others.[59]

Conclusion

The aim in this chapter has been to show that the use of unmanned systems can introduce a morally problematic asymmetry and that, at least in some cases, there are reasons for rejecting the legitimacy of unmanned engagements on grounds that stem from this asymmetry. Further, it has been argued that it is not so easy to justify an overwhelming asymmetric attack based on the supposed non-equality of combatants. A military with great technological power must thus think twice about the threat that their enemy poses. In considering whether radically asymmetric unmanned warfare really is the best option to achieve a military's desired goal, it must take a long-term and more strategic view of risk, considering all possible evoked responses. On the basis of the evidence and arguments put forward in the previous six chapters, it seems clear that there is a role for drones and other remote weaponry, since they have the potential to offer numerous benefits. However, several questions remain concerning their deployment. In relation to the problem posed by asymmetry, we must ask: in what capacity should unmanned weaponry be utilised, and to what degree? We must be careful not to miscalculate threat in the 'fog of war' and not to err too far (unjustifiably so) on the safe or dangerous side when faced with a challenge. This is especially true with regards to the political, psychological and time pressures which are often placed on key decision makers in unmanned warfare, as was discussed in chapters 5 and 6. Going forward, we need to think about the effect that asymmetry and other factors can have on enemy populations and the prospects for lasting peace.

Notes

1 David Rodin, 'The Ethics of Asymmetric War', in *The Ethics of War: Shared Problems in Different Traditions*, ed. Richard Sorabji and David Rodin (Aldershot: Ashgate, 2006), 58.

2 William Safire, *The Right Word in the Right Place at the Right Time: Wit and Wisdom from the Popular 'on Language' Column in the New York Times Magazine* (New York: Simon & Schuster, 2004), 13.

3 Ibid.

4 Andrew Mack, 'Why Big Nations Lose Small Wars: The Politics of Asymmetric Conflict', *World Politics: A Quarterly Journal of International Relations* 27, no. 2 (1975).

5 Ibid., 178.

6 Ibid., 184–6.

7 Safire, *The Right Word in the Right Place at the Right Time: Wit and Wisdom from the Popular 'on Language' Column in the New York Times Magazine*, 13.

8 The United States Joint Staff, *Joint Strategy Review* (Washington, DC: Department of Defense, 1999).

9 Safire, *The Right Word in the Right Place at the Right Time: Wit and Wisdom from the Popular 'on Language' Column in the New York Times Magazine*, 13.

10 Kenneth McKenzie, *The Revenge of the Melians: Asymmetric Threats and the Next Qdr* (Washington, DC: National Defense University, 2000).

11 Rodin, 'The Ethics of Asymmetric War', 155.

12 Enemark, *Armed Drones and the Ethics of War: Military Virtue in a Post-Heroic Age*, 59.

13 Ibid.

14 Steven Metz and Douglas V. Johnson, 'Asymmetry and U.S. Military Strategy: Definition, Background, and Strategic Concepts' (Washington, DC: United States Strategic Studies Institute, 2001), 6.

15 Mack, 'Why Big Nations Lose Small Wars: The Politics of Asymmetric Conflict', 182.

16 Juliet Barker, *Agincourt: The King, the Campaign, the Battle* (London: Little & Brown, 2005), 320.

17 Ibid.

18 Gary Taylor, ed. *Henry V* (Oxford: Oxford University Press, 1982). The sections immediately before and after the St. Crispin's Day Speech speak of the moral burden of being king, but the whole play can be taken as an anti-war allegory.

19 Daniel R. Headrick, *Power over Peoples: Technology, Environments, and Western Imperialism* (New Haven: Princeton University Press, 2010), 275.

20 Ibid.

21 Harold E. Raugh, *The Victorians at War, 1815–1914: An Encyclopedia of British Military History* (Santa Barbara: ABC-CLIO, 2004), 257.

22 Winston Churchill, *The River War: An Historical Account of the Reconquest of the Soudan* (London: Longmans, Green and Co., 1899). Based on his later actions as a politician, Churchill obviously thought that even dishonourable weapons had their legitimate roles.

23 Alastair Finlan, *The Gulf War of 1991* (New York: Rosen Publishing Group, 2008), 84.

24 Andrew Murphie and John Potts, *Culture and Technology* (London: Palgrave, 2003), 172.

25 Thomas G. Mahnken, *Technology and the American Way of War* (New York: Columbia University Press, 2008), 183.

26 Geoffrey Kemp, 'Arms Acquisition and Violence: Are Weapons or People the Cause of Conflict?', in *Leashing the Dogs of War: Conflict Management in a Divided World*, ed.

Chester A. Crocker, Fen Osler Hampson and Pamela R. Aall (Washington, DC: United States Institute of Peace Press, 2007), 60.

27 Adapted from Suzy Killmister, 'Remote Weaponry: The Ethical Implications', *Journal of Applied Philosophy* 25, no. 2 (2008).

28 Strawser, 'Moral Predators: The Duty to Employ Uninhabited Aerial Vehicles', 344.

29 Strawser actually levels three challenges at the 'problem of fairness'. I will look only first two as they contain his main argument, and the third is rendered false by the discussion of the first two.

30 Strawser, 'Moral Predators: The Duty to Employ Uninhabited Aerial Vehicles', 356.

31 Ibid.

32 McMahan, *Killing in War*. This recent monograph draws on many papers published over the last decade or so and represents some of McMahan's clearest thought on the topic.

33 See Walzer, *Just and Unjust Wars: A Moral Argument with Historical Illustrations*; Johnson, *Just War Tradition and the Restraint of War*; Temes, *The Just War*.

34 Walzer, *Just and Unjust Wars: A Moral Argument with Historical Illustrations*, 34–40.

35 McMahan, *Killing in War*, 15–31.

36 Michael Walzer, 'Response to McMahan's Paper', *Philosophia* 34, no. 1 (2006): 41–43.

37 Ibid., 43.

38 Paul Kahn, 'The Paradox of Riskless Warfare', *Philosophy and Public Policy Quarterly* 22, no. 3 (2002): 2.

39 Ibid., 3–4.

40 Sparrow, 'Robotic Weapons and the Future of War', 127.

41 Kahn, 'The Paradox of Riskless Warfare'.

42 Sparrow, 'Robotic Weapons and the Future of War', 128.

43 There is a parallel here with the utilitarian argument for torture: if we can torture someone to retrieve information that will save the lives of many people, without anyone finding out and thus setting a precedent for further tortures, then we should do so (under some accounts).

44 For a very interesting discussion concerning McMahan's 'deep morality' and its applicability in our non-ideal world, see Henry Shue, 'Do We Need a 'Morality of War'?', in *Just and Unjust Warriors: The Moral and Legal Status of Soldiers*, ed. David Rodin and Henry Shue (Oxford: Oxford University Press, 2008).

45 Enemark, *Armed Drones and the Ethics of War: Military Virtue in a Post-Heroic Age*, 60.

46 This reflects a more common problem concerning the projection of consequences within broadly consequentialist theories.

47 Walzer, *Just and Unjust Wars: A Moral Argument with Historical Illustrations*, 43.

48 What if the state does not feel compelled to act ethically? This might call for a different approach from State X, but the method to reach it is essentially the same: it involves considering potential responses and acting in accordance with these.

49 Killmister, 'Remote Weaponry: The Ethical Implications', 122.

50 Robert Simpson and Robert Saparrow, 'Nanotechnologically Enhanced Combat Systems: The Downside of Invulnerability', in *In Pursuit of Nanoethics*, ed. Bert Gordijn and Anthony Cutter (Dordrecht: Springer, 2013/14), 93–94.

51 Stephen Lendman, 'America's Drone Command Centers: Remote Warriors Operate Computer Keyboards and Joysticks', Global Research: Centre for Research on Globalisation, http://www.globalresearch.ca/america-s-drone-command-centers-remote-warriors-operate-computer-keyboards-and-joysticks/30590.

52 Chris O'Connell, 'Drone Command Center, 200-Plus Jobs Coming to Horsham', Fox News, http://www.myfoxphilly.com/story/21676061/drone-command-center-200-jobs-coming-to-horsham.

53 Andrew Croome, *Midnight Empire* (Sydney: Allen & Unwin, 2012).

54 Scott Wilson, Greg Miller, and Sari Horwitz, 'Boston Bombing Suspect Cites U.S. Wars as Motivation, Officials Say', *The Washington Post* (2013), http://articles.washingtonpost.com/2013-04-23/national/38751370_1_u-s-embassy-boston-marathon-bombings.

55 Stephen Nathanson, *Terrorism and the Ethics of War* (Cambridge: Cambridge University Press, 2010); Virginia Held, *How Terrorism Is Wrong: Morality and Political Violence* (Oxford: Oxford University Press, 2008); Michael O'Keefe and C. A. J. Coady, eds., *Terrorism and Justice: Moral Argument in a Threatened World* (Carlton: Melbourne University Press, 2002).

56 I am indebted to B. J. Strawser for raising this potential objection.

57 Rodin, 'The Ethics of Asymmetric War', 161–5.

58 Kahn, 'The Paradox of Riskless Warfare', 4–5.

59 For some ideas stemming from earlier drafts of this chapter, see Megan Braun and Daniel Brunstetter, 'Rethinking the Criterion for Assessing CIA-Targeted Killings: Drones, Proportionality and Jus Ad Vim', *Journal of Military Ethics* 12, no. 4 (2014).

Chapter 8
Unmanned Systems and War's End: Prospects for Lasting Peace

The majority of people probably remember US President George W. Bush alighting from a fighter jet and onto the deck of the aircraft carrier USS Abraham Lincoln to give a televised war speech under a gigantic banner reading 'Mission Accomplished'. In his May 2003 address, given to signal the supposed end of major combat operations in the Middle East, President Bush stated that the 'United States and [its] allies have prevailed'.[1] Bush's assertion, and the banner itself, continue to be quite controversial to this day. Indeed, it remains unclear whether this was the right message given that the vast majority of military and civilian casualties occurred long after his speech, despite the wide scale deployment of technologically sophisticated unmanned systems. While it is too simplistic to say that the United States-led coalition failed to plan for the aftermath of the wars in which it became involved, it can be submitted that the poor prospects for lasting peace must be partly attributed to the lack of attention given to the effect of unmanned warfare on the so-called 'third tenet' of just war theory. This tenet, *jus post bellum,* is concerned with whether the termination phase of war is just. It offers both an analysis and an opinion on the responsibilities of drone-wielding states in ensuring a smooth transition from conflict to peace. The first half of the chapter is largely theoretical and examines the historical and conceptual foundations of just war theory's third tenet. It maps current thinking as well as provides insights on the content of *jus post bellum* and its synergistic relationship with *jus ad bellum* (which governs responsibilities in resorting to war). The second half of the chapter bridges the gap between theory and practice. It begins by exploring the issues raised by the demands of *jus post bellum* in the context of the use of unmanned systems. This involves providing a general assessment of the role that distance and technology play in forming the perception of the war among an adversary's culturally diverse population. It also involves looking briefly at what engineers and strategists can do to overcome the conflicts between unmanned warfare and the ends of *jus post bellum.*

Historical Conceptions of a Tripartite Just War Theory

Nearly every discussion of unmanned systems pays homage to classical just war theory. According to the moral perspective adopted in Chapter 4, we know just war theory to be a coherent set of concepts and values designed to enable systematic moral judgement in wartime, with two main tenets: *jus ad bellum* and *jus in bello*. The principles within these categories are not unproblematic, given recent challenges to the division between the two tenets by people like McMahan. However, as demonstrated in Chapter 7, these challenges have not yet proved fatal. Just war theory continues to hold a great deal of explanatory power and many important ethical issues related to unmanned systems are fruitfully analysed under its main tenets. Yet it may be argued that relying on the above two tenets when assessing unmanned warfare does not do justice to the task and leaves out an important question, which is related to the nature of post-war responsibility. Accordingly, some just war theorists have argued that we need to develop just war theory. They say that it should include an additional tenet dealing with principles covering demilitarisation, compensation, rehabilitation and so on. This third tenet has been labelled *jus post bellum*. Given that the hostilities in Afghanistan and Iraq continue despite the end of actual combat operations, the withdrawal of troops and the many assurances that the focus is now on the reconstruction of these nations, there has been surprisingly little academic enquiry into this recent extension of just war theory and virtually no scholarship linking it to unmanned systems technologies.[2] Before examining this relationship in greater detail, attention needs to be given to the relevant, but limited, history of politico-moral and military thought concerning responsibilities after war's end, as well as why the typical just war theorist disregards or rejects this intellectual framework.

While some scholars regard *jus post bellum* thinking as being in its theoretical infancy,[3] it does have a place in the work of a number of influential just war scholars. Historical traces of a tripartite conception of just war theory can be found across many of the different traditions that have contributed to modern just war thought.[4] St Augustine linked war to the post-war goal of peace in his book *City of God*.[5] He pointed to the requirement to conduct war in a way that is conducive to long-term goals in writing that 'peace is the instinctive aim of all creatures, and is even the ultimate purpose of war … even in the extreme case when they have separated themselves from others by sedition, they cannot achieve their aim unless they maintain some sort of semblance of peace with their confederates in conspiracy'.[6] Much later, other proponents of just war theory built upon this thinking and the idea of a tripartite conception. These included Francisco de Vitoria, who set basic rules for the just seizure of the enemy's property and demanded that reparations be paid,[7] and Francisco Suárez, who gave support for similar rules when writing that a state has no reason to continue fighting where 'full and sufficient satisfaction is voluntarily offered'.[8] More recently, Hugo Grotius developed an explicitly tripartite conception in *On the Law of War and Peace*,[9] which secularised just war principles and held them to be binding on all states regardless of local custom. Book Three of that work, in particular, includes not only the typical rules that we have come to embrace in the form of *jus ad bellum* and *jus in bello*, but also practical rules on the just cessation of war, governing areas such as surrender and the peace treaties that typically follow.[10] Kant also put forward a theory based on three tenets: Right to War, Right in War and Right after War.[11] He associated the latter, his equivalent of *jus post bellum*, with justice as it is manifested in: peace settlements, the limits on the punishment of individuals and the respect for self-governance.[12] Thus, all of these just war theorists agreed, on some level, that states ought to be held responsible for the way in which war is ended and the prospects for peace managed.

Despite the work of these classical theorists, most modern debates concerning war and warfighting technologies end with little formal attention given to the ethics of the post-conflict environment. This does not suggest that the just war tradition is simplistic or lacks modern relevance. As argued elsewhere, the founding fathers of just war theory could not have realistically imagined what the problems posed by teleoperated unmanned machines would be like, let alone fully autonomous unmanned systems, but the main principles they suggest remain sturdy and thought-provoking when applied to changing political systems and battlespace innovations. The case is the same for *jus post bellum*; it is a matter of re-examining and re-applying it in novel contexts. However, as Orend has pointed out, it still needs to be asked why the termination phase of war has been overlooked in recent times and whether there is anything significant or revealing in this very fact of it being overlooked?[13] Here, mention is made of two of the more convincing reasons that may have been a factor in the scholarly disregard of *jus post bellum*.

The first relatively simple reason is that in earlier times, the ends of war were different, and so too were the motivations behind war. For instance, in the Middle Ages war was limited by the ideals of the church, as Augustine writes, and also by more realistic constraints. Historian James Turner Johnson writes that war was more of a part-time endeavour because of the challenges associated with fighting throughout much of the year, and because armies were drawing on the larger population base for their economic support. This meant that to avoid being destitute, warriors needed to oversee their properties, ensuring that they remain productive and self-sufficient. As a result, the situation for warriors was more like what it is today for Army reservists.[14] The weapons employed during these conflicts were also difficult to use and, in the case of trebuchets and battering rams, notoriously difficult to transport. Therefore, it was not in the interests of the belligerents to employ harsh tactics or destructive weapons. To do so would be to counter-productively threaten their political and economic wellbeing. With a strong regard for a peace and wellbeing already built into the means and nature of warfare, this eliminated much of the need for an explicit third tenet.

The second and somewhat deeper reason has to do with the inertia of the tradition. As critics often emphasise, just war theory began along two established lines of thought regarding *jus ad bello* and *jus in bello* and many have shown what Orend calls an 'unreflective unwillingness' to break out of these two categories.[15] Others say that, while *jus post bellum* should not be considered to be a distinct category of just war theory, it is not entirely separate from it either. Those who fall into this later group talk of *jus post bellum*, but make a conscious decision to subsume it under the well-established category *jus ad bellum*. Following in the footsteps of Aristotle and many other just war giants, Walzer also subordinates post-war justice to *jus ad bellum*. In the first edition of his classic *Just and Unjust Wars*, he writes that 'justice after war consists in achieving the just cause which justified the start of the war to begin with … If the just cause was self-defence from aggression, *jus post bellum* consists in defeating and repulsing the aggressor … Full stop and we're done'.[16] Much more recently, he has written that the idea here is twofold: 1) war can only be considered just if there is a strong likelihood of success and, for this, leaders must have some idea of what success looks like; and 2) right intention means that whatever is taken to constitute success must not only be possible, but also morally defensible.[17] Thus, for him, and others in the subsumption camp, arguments about the post-war environment are a crucial part of the arguments about going to war in the first instance. That is, for these scholars at least, post-war considerations belong entirely within the category of *jus ad bellum*. Given that many scholars do indeed learn their just war theory at Walzer's figurative knee, this could partially account for the lacunae. However, we cannot leave the debate at this.

Why does a possible tripartite conception deserve renewed attention as we progress into the twenty-first century? For many of those who remain unconvinced by the earlier dismissal of McMahan's revisionist just war theory, the explanation is likely to be that a reconsideration of the bipartite structure could serve a very particular systemic function. Recall that those who embrace McMahan's contemporary just war theory – if it can be called such – reject the idea that the categories are logically distinct. The benefit for proponents of McMahanian type ideas, as Stahn notes, is that *jus post bellum* might have a unifying force and 'establish a closer nexus between the justification and motive of the use of force and the corresponding responsibilities in the aftermath of intervention'.[18] Through encouraging international actors to consider the broader consequences of their actions, formal recognition of the rules and principles governing the post-conflict period may also stop them from divorcing matters of conduct from those of the resort to force, closing the gap between *jus ad bellum* and *jus in bello*. Bridging this gap goes some way towards aiding the cause of just war revisionists. Of course, the proponents of classical just war theory will still maintain that, while the categories may inform each other to some extent, the justice of peace should be evaluated independently of the justice of the war that preceded it. That is, for the classicists willing to consider the notion of a third tenet, the combination of a just conflict with an unjust peace and vice-versa cannot be ruled out. For even where a state goes to war in compliance with the principles of *jus ad bellum* and fights in accordance with *jus in bello*, it may then fail to meet the conditions of *jus post bellum* after the war has ended.

More generally, and arguably more importantly, a contemporary theory of *just post bellum* would fill a large normative gap, given that, at present, there is no thorough competing moral framework which serves to govern the transition from conflict to peace. Some elements of the transition may already be partly incorporated in *jus ad bellum*, but since war has a beginning, a middle and an end, and each deserves a through and systematic treatment, we need a third tenet of just war theory that covers responsibilities in the closing phase of conflict. It is needless to say that it is best when wars actually end and end well. The answer to the question 'why do we need a third tenet?' is evident in the context of recent military interventions and their aftermath. While it is obviously difficult to set objective standards by which to judge the end-result of any given war, and while we will forever be striving to improve war's aftermath, few full-scale wars seem to have ended 'well' or as well as they could have. We need only turn to history and the trail of conflicts that have ended and led to devastating and long-lasting post-war conditions, often to be reignited into even deadlier conflicts. Iraq is but one recent example that underlines the challenges faced by states after the conclusion of the conventional phase of military operations. As such, the careful articulation of a third tenet may mitigate the gap and smooth over the transition from conflict to a better and long-lasting state of peace.

Even the proponents of the subsumption camp must admit that there are a number of advantages and some heuristic value in making it explicit as a separate 'checkpoint' or consideration.[19] It would, if nothing else, serve as a more detailed reminder to military and political decision makers of the need to take the goal of peace and any potential settlement into account in the balance of consequences when resorting to war and carrying out individual operations. In doing so, as the next two chapters will highlight, they must have a robust plan for dealing with any detrimental impact of military operations on prospects for ensuring the prompt restoration and/or implementation of peace.

The Principles of *Jus Post Bellum*

Without intending to deny the noteworthy efforts of some contemporary philosophers to introduce the category of *jus post bellum* and to carefully articulate its principles, thinking on the subject is, to borrow Robert Royal's apt description, in its infancy and certainly not beyond adolescence.[20] There may be a growing consensus – stimulated by George Bush's speech on that memorable day in 2003 – that the end and aftermath of war deserves a comprehensive and systematic treatment, but there is no general consensus on a canon of *jus post bellum* principles or criteria. As concluded by Camilla Bosanquet, having attempted to sort through the debate, 'the *jus post bellum* bandwagon, if you will, has filled up with musicians who have yet to get on the same sheet of music'.[21] Of course, this is not to say that the cacophony can be ignored. If anything, it is a further sign of the importance of carefully articulating relevant principles. Most existing articulations seek to create a synergy between *jus ad bellum* and *jus post bellum* through adapting the former's well-established principles to post-war problems or otherwise derive principles from best international post-war practice and propositions put forward by legal scholars. In this sense, the philosophers concerned with this articulation are doing what philosophers have always done: conceptualising the ideal while grounding it very much in the real.

In further mapping their thinking throughout this section, it is helpful to identify two separate positions or accounts, though it must be acknowledged that many scholars' work rests somewhere between the two and that the considerations captured by *jus post bellum* are sometimes hard to grasp. There are those who envisage *jus post bellum* as setting restrictive criteria outlining what it is permissible and impermissible for victors to do once the active conflict is over. Alex J. Bellamy labels this the 'minimalist approach'.[22] Drawing on the work of just war theorists such as Vattel and Grotius, Bellamy claims that minimalists tend to view war as a method of rights vindication and argues that combatants are only entitled to wage war if their rights are impinged upon and then only until these rights are vindicated. As a consequence of this thinking, minimalists are of the opinion that victors are entitled to protect themselves, recover anything that may have been illegitimately taken and punish the perpetrators who violated their rights or those of others, such as in the case of genocide.[23] In essence, the argument here is that they can restore the state of affairs to that which existed prior to the conflict, but any other interference or action is morally unsanctioned. Bellamy also identifies a maximalist approach. It builds on the less stringent account just recounted and adds two propositions: one, that there is an in-built presumption against conflict in just war thinking; and two, that the victors acquire duties and responsibilities to the vanquished which go well beyond the requirement not to exact anything more from the conquered than what is deemed necessary to vindicate their rights. These responsibilities include, but are not necessarily limited to, formalising the quasi-judicial function of war through establishing military courts and holding war crimes trials, as well as taking other active measures to avoid future war by assisting in long-term governance and reconstruction efforts, as the US did with Japan and Germany and is now attempting to do in the Middle East.[24]

The maximalist approach is clearly the more normatively appealing. The reason is straightforward: the minimalist approach merely encourages restoration of the *status quo ante*. Those who embrace this approach hold that victors are entitled to use only as much violence as is necessary to vindicate their rights and to return them to a state of affairs equal or as near as possible to the state of affairs that existed prior to the outbreak of war. But, given that war broke out in those conditions, to cease efforts upon reaching the *status quo ante* is to leave open the possibility of future conflict. If we are to avoid returning to a state of conflict, meaningful action is required. This leads us to the work of Orend, who is perhaps the foremost scholar and advocate of *jus post bellum*.[25] He argues that the scope of its principles must extend to 'a more secure possession of our rights, both individual and collective'.[26] The aim of a just war, he explains, is the vindication of the fundamental rights of societies, exercised on behalf of their citizens. He goes on to say that a legitimate, just war observing community must do all that it can reasonably do to: gain and maintain the recognition of its own people and the international community of states; adhere to basic rules of international justice, most notably non-aggression; and satisfy the rights of its citizens, including those to security. As a consequence of his rights-talk, his account may, at first glance, appear minimalist; but in his discussion of what these requirements actually amount to, its maximalist nature becomes clear. This is because Orend's discussion of *jus post bellum* makes little mention of entitlements and, instead, focuses on obligations. He enumerates the following six principles according to which victors acquire a wide range of responsibilities beyond those required to restore the *status quo ante*: rights vindication, proportionality and publicity of the peace settlement, discrimination, punishment, compensation and rehabilitation.[27]

On the first count, he stipulates that a just settlement of a war should secure the basic rights that were violated and which justified the initial waging of war.[28] The relevant rights could include those to life, liberty, sovereignty and, more controversially, territory. Securing these rights must, for Orend, be the substantive goal of any morally legitimate peace settlement. On the second count, drawing on the classical just war principle of proportionality, he says that the peace settlement must be measured and reasonable, as well as publicly proclaimed.[29] The call for a public proclamation ensures that the victorious state very clearly communicates to the belligerent state that it has sincere intentions for a lasting post-war settlement, while the proportionality condition ensures that the victor does not utilise the settlement as an instrument of revenge, which would only lead to more instability. Drawing on yet another familiar principle, he says that a distinction needs to be made between the soldiers, statesmen and civilians of the defeated state involved in the peace settlement.[30]

The latter group of persons is generally considered immune from the punitive aspect of post-war measures. This civilian immunity seems to rule out the imposition of broad socio-economic sanctions such as reparation taxes and the like. The soldiers and statesmen, on the other hand, are far from immune. Orend's fourth principle states that these categories of persons from the vanquished state must be punished proportionally to the moral crimes they facilitated.[31] Proportional punishment serves as a powerful deterrent against future moral crimes and may spur change and rehabilitation on the part of the aggressor. Of course, soldiers from all sides can commit war crimes and must be held accountable. Through his fifth principle, Orend further claims that the defeated state may also owe financial restitution to the victor, though as already stated, it falls to the latter to determine how this fits with the requirements of proportionality and discrimination.[32] Finally, and most importantly, he states that the aggressor state may require rehabilitation and demilitarisation depending on the nature of the violence it waged and the reasons for this violence, along with some consideration of the threat posed in the absence of such action.[33]

According to Orend, a just peace must meet all of these general requirements.[34] In short, there must be a comprehensive ethical exit strategy in place, regardless of the burden it puts on the minds of policy planners and the resources of military commanders. Any significant non-adherence would constitute a violation of just war theory, possibly leaving the offender open to moral condemnation. That is because those principles just enumerated represent a fair 'checklist'[35] of the areas that must be addressed in waging war with drones or more traditional weapons. However, more needs to be said about the principles and their requirements. These are not inexpensive undertakings, in either financial or moral-political terms. As the Western coalition's actions in the Middle East have demonstrated, trying to accomplish anything resembling these goals can be a decades-long effort and contribute to hundreds of deaths and casualties that would have been avoided had *jus post bellum* requirements not been fulfilled, at least in the short-term. Here rests a morally significant tension between a state's immediate responsibility for its own people, defined under the social contract explained in Chapter 3, and its strategic responsibility to care for the wellbeing of those of the defeated state. When not applied with sensitivity or full understanding of what is required, well-intentioned actions can also result in cultural collisions that may never be truly healed. Part of the problem is that the principles are very general in nature. With that said, different wars entail different post-war responsibilities. In some instances, a quick withdrawal might be prudent, while in others it may be wise to engage in lengthy stability operations. With the benefit of hindsight, the aim in the following section is to provide a more detailed understanding of these principles as they apply to drone-dependent operations, using the Middle East as an illustrative example.

Jus Post Bellum and the Drone Effort

From the above discussion, we know that under Orend's account of *jus post bellum*, states need to adhere to six general principles which are thought to aid in the establishment of a minimally just society and secure a better state of peace. What military commanders need to know, however, is how to operationalise and apply these (and similar) principles in the conflict zone. Translating these theoretical principles into practice is a difficult task given that post-war actions must also take into consideration the political, social, cultural and religious norms of the area in which operations are conducted. This is a task that the United States failed to give adequate attention to when preparing for the intervention in the Middle East. Much more is now known about this intervention's strategic planning or lack thereof. For instance, we know that the US military was not prepared for the long and deadly conflict that eventuated. Its forces were equipped for a short conflict that the US believed would involve nuclear, biological or chemical weapons.[36] We also know that American post-conflict planning was limited to protecting oil infrastructure and avoiding Kuwaiti-style destruction. While there were some successes in the sense that full-scale combat operations ceased within two months and Iraq's oil fields survived relatively unscathed, sectarian violence continues today, and the relevant governments do not seem to be more just now than they were prior to intervention. More than a decade after the initiation of what was supposed to be a five month operation, the US, economically exhausted and suffering from greater than expected losses, has come to rely ever more heavily on drones. This, it will be argued, has exacerbated the problem rather than alleviate it. This section will outline how to ground Orend's *jus post bellum* principles and demonstrate that the implementation of these principles is in tension with the reality of how unmanned systems are usually employed.

In addition to his six *jus post bellum* principles, Orend provides a 10-point blueprint aimed at maximising prospects for lasting peace whilst forcefully enacting politico-military change. These will be drawn on to illustrate the problem with unmanned warfare. The first point he makes consists in saying that states must adhere diligently to the laws of war during the conflict and any subsequent period of occupation.[37] This is morally vital for its own sake, but also serves to reinforce a state's cause and the legitimacy of its occupation to the local population. The US has already run afoul of this principle in Iraq, owing to the notorious abuse and mistreatment of prisoners of war at Abu-Ghraid prison, but the focus has already started to shift from these concerns over torture and humiliation to those about drone warfare. As already highlighted throughout this book, the employment of unmanned systems can breach the rules of war for any number of reasons: it can lead to more wars, lower barriers to killing and be intrinsically unfair and unjust in cases of radical asymmetry.

However, perhaps the biggest concern is that these systems might be perceived as being indiscriminate and/or disproportionate. The history of the targeted killing facilitated by unmanned systems technologies is marked by a long series of errors that are said to have resulted in large numbers of civilian casualties and mass destruction. According to Plaw, while accounts of the imprecision of these systems differ, many report effective discrimination or 'kill rates' ranging from 10 per cent through to an abysmal two per cent; that is, according to some accounts, only one in 50 drone attacks are said to have resulted in the deaths of al-Qaeda combatants or other militant group operatives.[38] While it goes without saying that these claims and figures are often exaggerated, there is a common perception that Presidents Bush and Obama have lost the moral high ground as a result of the drone campaigns spanning their terms of office.[39] This is problematic, because if both the local population and the international audience begin to have doubts about a state's moral basis for continuing to wage war, they may withdraw their popular support and resources, and this will either undermine the state's chance of success altogether or extend the time it takes to accomplish a state's set goals, and so will lead to longer, bloodier wars.

Another consequence of conducting inaccurate and disproportionate drone strikes and allowing society to have misgivings about the moral rectitude of such operations is that it makes the disarmament and demilitarisation of a society during war and in post-conflict phases of operations inordinately difficult, when this is critical to the establishment of a just and lasting peace.[40] Many military commentators are concerned that unmanned systems technologies may actually be contributing to violence, hostility and instability in those areas where they are deployed.[41] Although some commentators disagree and highlight that there is no conclusive proof of a causal link between drone attacks and increased violence, there is strong circumstantial evidence, which suggests that the use of drones is spurring discontent and hampering the US efforts to quell the number of violent attacks that have been threatening their troops and others in their vicinity. For instance, Australian Lieutenant-Colonel David Killcullen, who is a former counterinsurgency advisor to General David Petraeus, has said that if the US continues with its aggressive drone attack policy, the tribespeople who are not militants will soon become extremists.[42] This may seem a bold claim, but imagine the equivalent domestic scenario. Imagine that you live in a rough American neighbourhood amongst drug lords, murderers and homeland terrorists. With the support of the state governor, the local police force decides to employ armed drones to capture and/or summarily kill those 'undesirables' living nearby. These criminals pose sufficient danger that the risk matrix put in place to weigh civilian harms against the benefit of eliminating these people from society warrants conducting operations in your high-density neighbourhood. The police accidently destroy a number of homes and a local park. Are the homeowners and families of the injured more likely to turn against the criminals or the police force? And if they wanted to apprehend the criminals themselves, how would they go about that?

We ought to apply the same sort of logic to unmanned warfare and grasp the idea that killing noncombatants or destroying their homes and murdering their loved ones means fostering conditions that are antithetical to the establishment of enduring peace and likely to give rise to radical insurgence. In both Iraq and Afghanistan, the local populace's will to resist American efforts continues to grow stronger with each civilian killed or injured, each village destroyed and each family that is forced to live on the streets.[43] The potential strategic implication is that each of those affected represents an extended and close-knit network of family, friends and professional colleagues, and people in this network will be reminded of Britain and Pakistan's earlier aerial bombardment campaigns. This will only help to give them an additional reason to hate the United States and Western ideology. As Kaltenhaler et al. have recognised, radical terrorist organisations gain moral legitimacy as a force fighting against what appears to some as the latest attempt at colonisation.[44] As one young independent fighter is reported to have told *The Guardian*, 'if young men lose hope in our cause they will be looking for an alternative. And our hopeless young men are joining al-Qaeda'.[45] This view is supported by a United Nations report, which concluded that air strikes were among the principal motivations for suicide bombers.[46] Furthermore, an American study, which consisted of interviews with Yemeni government officials, tribal leaders and others, concluded that the unintended consequence of drone attacks in Yemen has been a 'marked radicalisation of the local population'.[47] As Harry Bash put it, drones may be formidable weapons, but the Taliban and al-Qaeda have devised their own low-cost weapon in the form of 'suicidal jihadists who have impacts not unlike that of our aerial ones'.[48] To that end, they recruit disaffected youth, inculcate them with radical ideals, convince them of the need for some religiously inspired jihad and then have them wear explosive-ridden clothing, which is then detonated in some densely populated area. This is a similar point to that made about state-sanctioned warfare in the previous chapter: when one is left with no alternative option, some people will resort to whatever means are available to effect change. For Americans, there is a stark contrast between the American soldiers and the suicidal jihadists. But for many Iraqis, Afghanis and others affected, the difference between these two categories is likely to be much less significant.

The fabled precision of drone attacks also decreases the likelihood that non-state associations and post-war civil society will eventually flourish, which is another practical requirement under Orend's *jus post bellum* blueprint. A recent report from Columbia Law School on drone warfare identifies a number of relevant factors that make the affected public reluctant to associate with others, enjoy life and show a commitment to their society. First, a number of neutral civilians have found themselves caught in a precarious and rather dangerous position between local militants belonging to groups such as al-Qaeda and the United States drone effort. They have been suspected of being spies or informants working for the foreign coalition.[49] In one incident reported by the *Los Angeles Times*, a shop keeper was abducted, thrown into the back of a car and taken to a safe house where, along with other suspected 'traitors', and he was beaten for eight weeks.[50] Unable to determine whether he was in fact guilty, he was released to another militant group and later set free. While those suspected of colluding with foreign powers have been harmed and present the main concern, many of those who have been mistakenly targeted by drone missiles are also stigmatised and assumed to be connected in some way to the militancy.[51] In such cases, civilians living nearby those liable to be bombed are likely to undertake their own vigilante action against the suspected militants in the hope of ridding their community of potential danger, not knowing that apart from having a particular affinity for hitting large gatherings such as weddings and funerals, drones have also been known to mistake the homes of peace activists for Taliban hideouts. This worry about unmanned systems and negative association is of particular moral concern because they may inhibit or discourage the formation of civic groupings that connect citizens to their government (and often protect them from the same), take pressure off the state and serve to legitimise the state. Without the sort of robust civil societies that strong public participation supports, tyrannical governments can take reign and lead to further violence and war.

To prevent the rise or re-growth of tyrannical governments, Orend proposes that the victorious state must work with a cross-section of the local population on a new constitution and minimally just government. The latter must incorporate checks and balances to ensure that those whom it serves are protected as well as possible.[52] In terms of the constitution building process, he is careful to say that all those whose non-participation could ruin any subsequent agreement, and anyone who is committed to a minimally just state, must be included in some fashion.[53] The picture that Orend paints is one of genuine and well-intentioned political partnership between the victor, the vanquished and its population; and while it is obvious why the support of all three is necessary, the employment of unmanned systems threatens the meaningful participation of the latter for much the same reasons mentioned above. Apart from the abovementioned, the unwillingness to cooperate with American forces and the transitional government initially implemented in Iraq can surely be attributed, more generally, to a failure to secure the oft-cited hearts and minds of locals.[54] Securing the emotional and intellectual support of this populace is problematic for an unmanned military force, for it is clearly imaginable that sending out robots to fight one's wars and kill one's enemy does not sit well with those on the ground. Enemy combatants might take it to be indicative of the fact that their counterparts are not willing to settle their dispute in the customary, and somewhat more chivalric, face-to-face manner. It is known for a fact that the Taliban, accustomed to fighting in close proximity to their Middle Eastern enemies, sees the use of these distance weapons as extremely cowardly.[55] This may be the same for civilians too, for as already explained, they may only ever hear the powerfully alienating hum of the Predator drones flying overhead and see the destruction and side effects they cause, without ever seeing the faces of those operators behind them.[56] Furthermore, there are far fewer, if any, soldiers on the ground to take responsibility for any collateral damage and carry forward values like honour, honesty, duty, respect, integrity and personal courage, which are traditionally attributed to warriors who are considered to be representatives of their respective states. However, the Middle East has not diverged from traditional practice as quickly as the West, so it is precisely these sorts of values that the US needs if they are to win the approval of Iraqis, Afghanis and other national populations.

Unfortunately, in many respects, the drone war across Iraq, Afghanistan and its borderlands with Pakistan invite what is probably an unwelcome comparison to the Vietnam War.[57] While the Obama administration and senior officials in the Pentagon may not be perturbed, the use of drone aircraft to rain bullets and launch missiles at key targets in what are often urban areas has become nearly as ritualised as the sustained bombing campaign against North Vietnam, which was later code-named 'Rolling Thunder'. In both cases, the US has displayed a reluctance to commit ground troops and lost the support of the local population due, at least in part, to the means employed. Levels of popular support are equally concerning. In their working paper on Pakistani public attitudes towards drone strikes, Kaltenhaler et al. cite the (2010) Pew Global Attitudes Project poll that confirms the reasons for the loss of support for unmanned warfare in that country. It found that, while the United States was not universally hated, the wide majority opposed the drone attacks the US conducted in conjunction with their government.[58] When asked whether drones kill too many innocent people, very strong responses were elicited from those who answered, with 95 per cent of its 2,000 respondents answering in the affirmative.[59] Figures from the most recent (2012) Global Attitudes Project poll confirm slightly lower disapproval levels, but these remained constant across the populations of 17 of all 20 states surveyed.[60] There is not yet any evidence that this is a higher level of opposition and a perceived threat that would flow from non-drone warfare, but Kaltenahler et al. argue that much of the opposition, is, in fact, driven by an adversarial image of the US presented by militant groups and religious extremists.[61] It must be said both that this alone cannot account for the difficulty in securing the approval necessary to avoid post-war rumblings, and that religious and political propaganda cannot really be separated from the true underlying causes already discussed. Yet, it must also be taken into consideration that if the US and those who will inevitably follow in its footsteps in employing drones are to forge working relationships with those they have conquered, they must provide help to develop the instruments of minimally just society in the aftermath of a conflict.

This leads to the penultimate point and one of Orend's final requirements, which stipulates that, in order to avoid a return to the previous state of affairs, the states must purge the past regime's propaganda or the militants who had infiltrated the target state, and they must encourage a shift in shared national values.[62] He suggests that this might necessitate revising the educational curriculum in the affected countries or provinces since, in the early twentieth century, some fascist groups used the school system to indoctrinate children, i.e. future citizens, so that they would be more inclined to subscribe to their doctrines of racial and national supremacy.[63] In the past, this has resulted in a great deal of aggression on the part of the indoctrinated. The same holds today, but it is the teaching of Islamic extremism – with its attitudes towards violence and women – and the employment of unmanned systems that often fuel rage amongst the locals and create the perfect environment for disseminating damaging disinformation and political propaganda.[64] Take al-Qaeda, which likes to be seen as trying to unite the Islamic world under a caliph by removing the presence of the US and pro-Western regimes from the world. It is popular across the Middle East because it has started to fill the void left by the governments of the countries in which it operates, in some cases caring for the sick and wounded and providing education for youth in remote areas. It is also quick to capitalise on drone strikes and surveillance by highlighting these practices in its propaganda. For instance, in 2009 it labelled its bombing of the Central Intelligence Agency's base in Khost as an act of 'revenge' for the deaths of Pakistani militants. Shortly after the trainer responsible for facilitating the attack was killed, his 'martyrdom' was boasted on jihadist websites and chatrooms.[65]

To completely rid society of this sort of propaganda and increase both states' chances of securing a less violent future, the victor must assist the vanquished state to establish reliable authorities to rid its new regime/administration of anyone materially connected with tyranny and atrocity, as well as capture and prosecute those amongst them who are likely to have committed war crimes (these are typically those also behind the propaganda effort).[66] This is, of course, important in its own right, since it ensures that the guilty receive their just deserts and that there is some force in operation to deter would-be terrorists or tyrannical dictators. Sadly, while some American unmanned aircraft do have very impressive surveillance capabilities that one would think could locate and track these individuals, they are technically limited as described in Chapter 5. Many militant groups that foster these individuals, now accustomed to fighting in drone-saturated environments, are also aware of these limitations and have developed tactics for evading drone surveillance. This is confirmed by several documents found by the Associated Press in buildings recently occupied by al-Qaeda fighters. Some documents were propaganda seeking to undermine America's drone effort by describing it as a last-ditch effort. Others were giving the fighters tactics to evade drone surveillance by telling them to stay in places unlit by the sun, to avoid gathering in open areas, to form decoy gatherings and possibly to seek to jam, confuse or override operating frequencies, among other things.[67] From information released about the eventual capture of Osama bin Laden, it is known that he too knew of these tactics, never venturing outside his compound in Abbottabad and never once looking up, which would have rendered him vulnerable to facial recognition. Finding such individuals is, of course, key to prosecuting them. When we consider unmanned systems in the context of the US 'capture or kill' policy, they are clearly better at the latter. The targeted killing facilitated by drones may be expedient for a political leader who disdains detention and interrogation as a matter of strategy, but it is not always the best choice in the war for gaining the approval of the population and fostering an enduring peace. This is particularly so when you consider that deadly strikes launched from unmanned systems remove the opportunity to extract valuable intelligence from those behind the violence.

Of course, every weapon used in combat operations – whether it be infantrymen, nuclear bombs, fighter jets or drones – has some sort of tactical, strategic and moral cost. In the case of unmanned aircraft, most of these costs are to some degree offset by the benefits they have over other weapons, namely a morally significant enhancement of capability and reduction in risk to otherwise vulnerable military personnel, as explained in Chapter 3. However, the current drone program challenges *jus post bellum* and the very thing that everyone hopes underpins the unmanned warfare campaign: the desire for a more secure possession of rights and freedom and a long-lasting peace for all involved. That said, there is hope that drones can actually be used to aid *post bellum* operations.

Drones for Good: A Note on Overcoming the Stigma

To summarise, *jus post bellum* requires that we fight war in such a way that the door is left open to lasting peace, but drones challenge the application of some *jus post bellum* principles, specifically practical requirements. Some ethicists hold that the challenges presented by drones are insurmountable. Sparrow, for instance, says that there is more to war than simply 'blowing people up' and that it is very clear that the Predator drone is precisely the wrong weapon to use to win the war in the Middle East.[68] According to his logic, if the goal of the war does indeed involve winning the approval of the locals and creating conditions that would allow them to govern themselves as well as live without the violence that has plagued them for decades, destroying key targets from drones flying high up in the sky will never accomplish this for many of the reasons noted above.[69] While we cannot deny the extent of the challenge in overcoming the drone stigma that has surfaced as a result of what has historically been offensive use, this is overly pessimistic. As Sparrow himself notes, there is more to war than killing people and drones need not be used exclusively for this purpose. Interestingly, it is in peace and stability operations that the US and others employing these systems have an opportunity to overcome the stigma attached to unmanned systems by demonstrating that they can play an active role in assessing and responding to the needs of peace.

In terms of fulfilling *jus post bellum* obligations, the US must first ensure that it correctly assesses the civilian harm caused by its drone attacks during active combat operations. In conventional military operations, military protocol calls for an enemy 'battle damage assessment', updated during conflict and completed at the end of any lethal operation.[70] The purpose is to determine whether objectives were met and the overall outcome accomplished, but, more importantly for our sake here, to assess whether any civilian harm may have occurred during the process. If it is suspected that civilian harm has resulted, an investigation will be conducted to confirm losses, learn practical lessons to prevent future harm and, in many cases, compensate those affected by the losses with monetary payments, counselling or other assistance. It is important to transparently take this basic step in the typically covert drone context for a number of reasons. First, it is important to assess and investigate civilian harm, as it is not only enemy personnel who should be prosecuted for war crimes or causing excessive collateral damage. Second, from a more strategic standpoint, a thorough assessment based on whatever information is attainable from in-weapon cameras may offer a counter to false allegations of civilian harm and is critical from an operational perspective. Third and finally, promoting transparency might send a clear and positive message to those in the Middle East affected by drone operations that the United States is committed to preserving human life and dignity.[71] At present, with the secrecy surrounding America's drone war and the government's reluctance to even acknowledge its existence, let alone its moral or legal basis, much of the local populace is angry, scared and prone to doubting the legitimacy of each resulting fatality rather than cooperative.[72]

The US and others can, however, utilise unmanned systems technologies in future conflicts to enhance their public affairs mission and communicate their intentions and policies to indigenous populations throughout the spectrum of military operations, minimising civilian repercussions in the *post bellum* phase and further countering any damaging disinformation that has been spread by the enemy. For instance, in remote areas of Iraq or Afghanistan, drones could be used throughout the spectrum of military operations to distribute messages written in the local dialect, explaining what these technologies are and why they are being used over local neighbourhoods. They could also be used to rapidly replace communications infrastructure damaged in battle in order to communicate with the local population via radio or mobile telephone. The Swarming Micro Air Vehicle Project, for example, utilises a dynamic swarm of high-flying drones that create a roving network capable of providing wireless communications over areas where the terrain would otherwise prevent such communication.[73] This would be particularly advantageous for those in mountainous areas of Afghanistan. This network might also be used to communicate truthful and factual unclassified information to the domestic population about the progress made towards achieving whatever goal initiated the conflict. All of these are elements of a successful *jus post bellum* programme. Therefore, drones can be used for good and fulfil a crucial and complementary function in presenting the other side of the story and shaping what intelligence officers specialising in psychological operations would call the 'cognitive domain' of the target population/s.[74]

Furthermore, unmanned technologies can be used for good as an explicit part of post-conflict rehabilitation and stability operations. After war, those who live in remote locations and are in need of assistance as a direct or indirect result of military operations are often the last to be attended to, because these areas are difficult to access for the troops or aid workers. Despite the fact that corrupt officials regularly stem the flow of humanitarian aid into these areas, organisations such as the United Nations and the Red Crescent movement are reluctant to put their personnel at risk by sending them into zones like these, regularly controlled by militant groups and hostile to external intervention in local affairs. It is in this regard that drones have a purpose beyond their strictly military origins.[75] As just one example, large unmanned aerial vehicles could be used to supplement or restore existing supply lines and provide for emergency drops. NATO, struggling with serious logistics problems in high-threat areas of southern Afghanistan, has already evaluated the use of pilotless helicopters for exactly this purpose. The K-MAX task-autonomous helicopter was tested over a period of six months and it has successfully delivered several hundred tonnes of cargo since its inaugural flight.[76] These robotic helicopters eliminate the need for convoys and large numbers of ground crew who would be at risk if they entered the field. Employed more creatively, unmanned systems could also be used as an ongoing measure to detect large groupings or other stirrings that have the potential to return states to conflict, though this sounds much like 'uberveillance' and could raise privacy concerns.[77]

It should be remarked here that this final point differs from that made by Savulescu and Beauchamp in their paper on teleoperated combat vehicles as 'robot guardians', discussed earlier.[78] Savulescu and Beauchamp are drawing attention to the fact that, insofar as drones limit harm to soldiers and aid workers, these systems may make states less concerned about incurring casualties and more likely to engage in humanitarian military interventions, which they suggest is a good thing. Again, there is a reason why one should not be convinced by this argument. While states do sometimes fail to intervene when they ought, they have a far greater tendency to intervene when it is either unnecessary or unwarranted, only to make the state of affairs significantly worse. For this reason if no other, it is important to differentiate the two arguments. It also needs to be said that there are some clear obstacles to employing drones in support of post-conflict aid operations and overcoming the stigma that has become cemented in the minds of the Middle Easterners. Most of these are related to the fact that the dual-use nature of unmanned systems technologies will always leave room for uncertainty about the true nature and intent of drone operations, for even the best uses can surreptitiously aid paramilitary, military or state endeavours. The decision to use unmanned systems must, therefore, be tempered by these and other *jus in bello* considerations.

Conclusion

The wars in the Middle East will be the defining conflicts of our time once all related operations draw to a genuine close. They have raised a myriad of issues, from those related to pre-emption and prevention through to those concerning sovereignty and independence. However, this chapter has argued that one of the most fundamental and overarching issues facing us today relates to the obligations and responsibilities concerning post-war reconstruction, stability and peace. The first section highlighted that, while in its modern theoretical infancy, *jus post bellum* has its place in the thinking of theorists such as Vitoria, Grotius and Kant, all of whom gave textual evidence of tripartite conceptions of just war theory. The second section detailed Orend's contemporary attempt at reviving the tripartite conception, exploring his six proposed principles. The third section bridged the gap between theory and practice. It has shown that unmanned warfare presents numerous challenges for Orend's 10-point blueprint for enacting these principles. Apart from these systems challenging the moral rules of war, unmanned systems make it significantly more difficult to disarm and demilitarise society, threaten non-state associations and the flourishing of post-war society, hinder efforts to work with locals towards establishing instruments facilitating lasting-peace, hamper efforts to purge harmful propaganda, and harm attempts to capture those suspected of war crimes. The fourth section acknowledged that there are ways to ensure that the drone stigma is, where possible, offset. Overall, this chapter has illustrated that unmanned aircraft can potentially play a critical, but often overlooked role, in hampering war's end and limiting prospects for lasting peace, if not properly managed. This is something the public, politicians, military leaders and engineers must be aware of as weapons development progresses and other machines that further exacerbate the problem are unleashed. With this in mind, we must consider who is responsible for any failure to take action in relation to this concern and the other potential problems.

Notes

1 Associated Press, 'Text of Bush Speech', CBS News, http://www.cbsnews.com/stories/2003/05/01/iraq/main551946.shtml.

2 Many scholars simply dismiss *jus post bellum* scholarship entirely, while others simply disregard it, because it complicates a neat discussion of the problems posed for *jus ad bellum* and *jus in bello*. For example, Daniel Brunstetter and Megan Braun disregard this important area of scholarship in one sentence, stating that 'finally, while such scholars as Brian Orend and Michael Walzer have argued for the importance of deepening our understanding of *jus post bellum*, we do not discuss it here'. See Brunstetter and Braun, 'The Implications of Drones on the Just War Tradition', 340.

3 Robert Royal, 'In My Beginning Is My End', in *Ethics Beyond War's End*, ed. Eric D. Patterson (Washington, DC: Georgetown, 2012), 65.

4 Carsten Stahn, 'Jus Post Bellum: Mapping the Discipline(s)', in *Just Post Bellum: Towards a Law of Transition from Conflict to Peace*, ed. Carsten Stahn and Jann K. Kleffner (The Hague: TMC Aasser Press, 2008), 95.

5 Augustine, *City of God.*

6 Ibid., 866.

7 Francisco de Victoria, 'De Indis Et De Ivre Belli Relectiones', ed. James Brown Scott and Ernest Nys, trans. John Pawley Bate, *The Classics of International Law* (Washington, DC: Institute of International Law, 1917), 184–85.

8 Francisco Suárez, 'The Three Theological Virtues', ed. James Brown Scott, trans. Gladys L. Williams, *The Classics of International Law* (Washington, DC: Institute of International Law, 1944), 838–40.

9 Hugo Grotius, 'On the Law of War and Peace: Three Books', ed. James B. Scott, trans. Francis W. Kelsey, *The Classics of International Law* (Washington, DC: Carnegeie Institution of Washington, 1925).

10 Ibid., 804–31.

11 Immanuel Kant, *The Philosophy of Law: An Exposition of the Fundamental Principles of Jurisprudence as the Science of Right*, trans. W. Hastie (Edinburgh: T. & T. Clark, 1887).

12 Ibid., 221–2.

13 Brian Orend, 'Jus Post Bellum: A Just War Theory Perspective', in *Just Post Bellum: Towards a Law of Transition from Conflict to Peace*, ed. Stahn Carsten and Jann K. Kleffner (The Hague: TMC Aasser Press, 2008), 33.

14 James T. Johnson, 'Moral Responsibility after Conflict', in *Ethics Beyond War's End*, ed. Eric D. Patterson (Washington, DC: Georgetown University Press, 2012), 23.

15 Orend, 'Jus Post Bellum: A Just War Theory Perspective', 33.

16 This is stated in the opening section to a chapter dealing specifically with *jus ad bellum*. See Michael Walzer, *Just and Unjust Wars: A Moral Argument with Historical Illustrations* (New York: Basic Books, 1977), 109–26.

17 Michael Walzer, 'The Aftermath of War', in *Ethics Beyond War's End*, ed. Eric D. Patterson (Washington, DC: Georgetown University Press, 2012), 35.

18 Stahn, 'Jus Post Bellum: Mapping the Discipline(s)', 321.

19 David Fisher, *Morality and War: Can War Be Just in the Twenty-First Century?* (Oxford: Oxford University Press, 2011), 79–80.

20 Royal, 'In My Beginning Is My End', 90.

21 Camilla Bosanquet, 'The Ethics of a Just Post Bellum', in *International Society of Military Ethics* (2007).

22 Alex J. Bellamy, 'The Responsibilities of Victory: Jus Post Bellum and the Just War', *Review of International Studies* 34, no. 4 (2008): 602.

23 Ibid., 603–5.

24 Ibid., 612–18.

25 Tony Coady puts forward a very similar elucidation of *jus post bellum* principles; however, in some sense it is even more comprehensive. His *jus post bellum* is all-encompassing, covering the termination of war, how to deal with post-war conditions and how to deal with one's own troops justly after the war. See C. A. J. Coady, 'The Jus Post Bellum', in *New Wars and New Soldiers: Military Ethics in the Contemporary World*, ed. Paolo Tripodi and Jessica Wolfendale (Farnham: Ashgate, 2011). As his account seems to conflate the subtle differences between these three sub-areas, Orend's account is utilised for the sake of this chapter.

26 Orend, 'Jus Post Bellum: A Just War Theory Perspective', 39–40.

27 Ibid., 41–42.

28 Ibid.

29 Ibid.

30 Ibid.

31 Ibid.

32 Ibid.

33 Ibid.

34 Ibid., 43.

35 The scare quotes are used here to avoid advocating the 'checklist approach', whereby our leaders and combat personnel come to view the ethics of war as a 'tick and flick' affair.

36 Eric D. Patterson, *Ending Wars Well: Order, Justice, and Conciliation in Contemporary Post-Confict* (New Haven: Yale University Press, 2012), 3.

37 Orend, 'Jus Post Bellum: A Just War Theory Perspective', 46.

38 Plaw, 'Counting the Dead: The Proportionality of Predation in Pakistan', 2–3.

39 Sarah Kreps, 'Flying under the Radar: A Study of Public Attitudes Towards Unmanned Aerial Vehicles', *Research & Politics* 1, no. 1 (2014). See, in particular, the public opinion data on support for drone strikes, located on page 3.

40 Orend, 'Jus Post Bellum: A Just War Theory Perspective', 46.

41 Avery Plaw et al., 'Do Drone Attacks Do More Harm Than Good', *The New York Times* (2012), http://www.nytimes.com/roomfordebate/2012/09/25/do-drone-attacks-do-more-harm-than-good.

42 Center for Civilians in Conflict, 'The Civilian Impact of Drones: Unexamined Costs, Unanswered Questions' (New York: Columbia Law School Human Rights Clinic, 2012), 23.

43 Leila Hudson, Colin Owens, and Matt Flannes, 'Drone Warfare: Blowback from the American Way of War', *Middle East Policy* 18, no. 3 (2011).

44 Karl Kaltenhaler, William Miller and Christine Fair, 'The Drone War: Pakistani Public Attitudes toward American Drone Strikes in Pakistan', in *Annual Meeting of the Midwest Political Science Association* (Chicago, 2012).

45 Ghaith Abdul-Ahad, 'Yemenis Choose Jihad over Iranian Support', *The Guardian* (2012), http://www.guardian.co.uk/world/2012/may/10/yemen-jihad-iran-saudi-interference.

46 Bergen and Tiedemann 2007, cited in Jeffrey A. Sluka, 'Drones in the Tribal Zone: Virtual and Losing Hearts and Minds in in the Af-Pak War', in *War, Technology, Anthropology*, ed. Koen Stroeken (New York: Berghahn Books, 2012).

47 Sudarsan Raghavan, 'In Yemen, US Airstrikes Breed Anger, and Sympathy for Al Qaeda', http://articles.washingtonpost.com/2012-05-29/world/35456187_1_aqap-drone-strikes-qaeda.

48 Harry H. Bash, *Stuff 'N' Nonsense: A Hodge-Podge of Whimsical, Sober & Politically Incorrect Musings* (Toronto: Inkwater Press, 2011), 461.

49 Center for Civilians in Conflict, 'The Civilian Impact of Drones: Unexamined Costs, Unanswered Questions', 22.

50 Alex Rodriguez, 'Pakistani Death Squads Go after Informants to US Drone Program', http://articles.latimes.com/2011/dec/28/world/la-fg-pakistan-death-squads-20111228.

51 Center for Civilians in Conflict, 'The Civilian Impact of Drones: Unexamined Costs, Unanswered Questions', 22.

52 Orend, 'Jus Post Bellum: A Just War Theory Perspective', 47.

53 Ibid.

54 Numerous scholars make mention of this. For two good discussions, see Joseph Cummins, *Why Some Wars Never End: The Stories of the Longest Conflicts in History* (Massachusetts: Fair Winds Press, 2010) and Antonius C. G. M. Robben, 'Losing Hearts and Minds in the War on Terrorism', in *Iraq at a Distance: What Anthropologists Can Teach Us About the War*, ed. Antonius C. G. M. Robben (Philadelphia: University of Pennsylvania Press, 2010).

55 Loes van Wifferen, 'Alienation from the Battlefield: Ethical Considerations Concerning Remote Controlled Military Robotics' (Universiteit Utrecht, 2011).

56 Center for Civilians in Conflict, 'The Civilian Impact of Drones: Unexamined Costs, Unanswered Questions', 23.

57 Lloyd Gardner, 'America and the Vietnam War: Re-Examining the Culture and History of a Generation', in *Iraq as the 'Good War' as Opposed to Vietnam, the Bad War*, ed. Andrew A. Wiest, Mary K. Barbier, and Glenn Robins (New York: Taylor & Francis, 2009), 309.

58 Kaltenhaler, Miller, and Fair, 'The Drone War: Pakistani Public Attitudes toward American Drone Strikes in Pakistan', 2–3.

59 Ibid., 4.

60 Pew Research, 'Drone Strikes Widely Opposed', http://www.pewglobal.org/2012/06/13/global-opinion-of-obama-slips-international-policies-faulted/.

61 Kaltenhaler, Miller and Fair, 'The Drone War: Pakistani Public Attitudes toward American Drone Strikes in Pakistan', 6.

62 Orend, 'Jus Post Bellum: A Just War Theory Perspective', 48.

63 Ibid.

64 Fancois Gere, 'The Future of Asymmetric Warfare', in *The Ashgate Research Companion to War: Origins and Prevention*, ed. Hall Gardner and Oleg Kobtzefff (Farnham: Ashgate, 2012), 522.

65 Hillel Ofek, 'The Tortured Logic of Obama's Drone War', *The New Atlantis: A Journal of Technology and Society* 27, no. 1 (2010): 38.

66 Orend, 'Jus Post Bellum: A Just War Theory Perspective', 46.

67 Abdullah bin Mohammed, 'The Al-Qaida Papers – Drones' (2011), http://hosted.ap.org/specials/interactives/_international/_pdfs/al-qaida-papers-drones.pdf.

68 Robert Sparrow, 'The Ethical Challenges of Military Robots', in *Ethical and Legal Aspects of Unmanned Systems: Interviews*, ed. Gerhard Dabringer (Vienna: Institute for Religion and Peace, 2010), 89.

69 Ibid.

70 United States Army, 'Field Manual' (Washington, DC: Department of Defense, 1994).

71 Center for Civilians in Conflict, 'The Civilian Impact of Drones: Unexamined Costs, Unanswered Questions', 61.

72 Ibid., 19–35.

73 Sabine Hauert et al., 'The Swarming Micro Air Vehicle Network (SMAVNET) Project', http://lis2.epfl.ch/CompletedResearchProjects/SwarmingMAVs/.

74 Joseph R. Didziulis, 'Winning the Battle for Hearts and Minds: Operationalizing Cultural Awareness During Stability Operations' (Montgomery: Air Command and Staff College, 2008), 9.

75 Jack C. Chow, 'The Case for Humanitarian Drones', *Open Canada* (2012), http://opencanada.org/features/the-think-tank/essays/the-case-for-humanitarian-drones/.

76 Isaac Lamberth, 'K-Max – Unmanned Aerial Alternative to Convoys', http://www.isaf.nato.int/article/news/k-max-unmanned-aerial-alternative-to-convoys.html.

77 M. G. Michael and K. Michael, 'Towards a State of Uberveillance', *IEEE Technology and Society* 29, no. 2 (2010).

78 Beauchamp and Savulescu, 'Robot Guardians: Teleoperated Combat Vehicles in Humanitarian Military Intervention'.

Chapter 9
The Responsibility Gap

Central to the ethical concerns raised about the development of increasingly intelligent unmanned systems are issues of responsibility and accountability. Robot arms control groups have popularised this element of the debate as part of their call for a moratorium on the use of autonomous drones. The purpose of this chapter is to demonstrate that, while unmanned systems certainly exacerbate some traditional problems and may in some cases cause us to rethink who we ought to hold morally responsible for military war crimes, our standard conceptions of responsibility are capable of dealing with the supposed 'responsibility gap' – namely the inability to identify an appropriate locus of responsibility – in unmanned warfare and that in the absence of a gap, there is no reason for an outright ban. This chapter begins by exploring the basis for the attribution of responsibility in just war theory. It then looks at the conditions under which responsibility is typically attributed to humans and how these responsibility requirements are challenged in technologically mediated warfare. Following this is an examination of Sparrow's notion of the 'responsibility gap' as it pertains to the potential deployment of fully autonomous weapons systems. It is argued that we can reach a solution by shifting to a forward-looking and functional sense of responsibility, which incorporates institutional agents and ensures that the human role in both engineering and releasing these systems is never overlooked.

Responsibility and Just War

In *Just and Unjust Wars*,[1] Walzer writes that the 'assignment of responsibility is … critical' because 'there can be no justice in war if there are not, ultimately, responsible men and women'. This is a logical claim as any defensible moral theory ought to incorporate a theory of moral responsibility or call for people to explain their intentions and beliefs if it is to lead to the rectification of wrongs. However, many theorists have simply asserted the importance of the assignment of responsibility in just war theory and proceeded to argue that the challenges presented for this assignment to intelligent systems (which can autonomously apply lethal force) are insurmountable, and so such systems should therefore be banned.[2] Sparrow, for instance, argues that it is a fundamental condition of fighting a just war that 'someone' be held responsible for any civilian deaths or other unjust actions.[3] He goes on to say that the assumption or allocation of responsibility is also 'vital in order for the principles of *jus in bello* to take hold at all'.[4] Lambèr Royakkers and Rinnie van Est take Sparrow's argument to mean that it is a condition of just war theory that an 'individual' be able to be held responsible.[5] One strong implication of the justice of warfare being separate from the justice of the war itself is that it permits the judging of acts within war to be disassociated from war's cause, even if all ties between cause and justice in war are not completely severed.[6] This is a useful division when it comes to examining Sparrow's former claim about the merit of *jus in bello* principles, though we first need to clarify the extent to which responsibility and accountability are necessary conceptual conditions for systematic compliance with the moral laws and customs of war.

When we speak of the necessity of attributing responsibility in this context, we are speaking of the need for accountability for harm or death caused in the targeting process and the subsequent decision to release (or not to release) deadly munitions. We know from just war theory that it is considered impermissible to attack people indiscriminately since non-combatants are considered outside the bounds of the battlefield. Their immunity in war stems from the fact that their existence and activity has little to do with war, with combatant status typically reserved for those involved in the prevention of harm and the rectification of wrongs through killing as agents acting on behalf of their respective states. Whoever or whatever is tasked with making targeting decisions must therefore proffer a reason as to why combatants become appropriate targets in the first instance. The difficulty of this task varies depending on the standards to which war making agents are held. Theorists have traditionally invoked rights analyses in order to give soldiers an idea of who is an appropriate target, though we will not go into these here. To build again on what was said earlier, some also hold that simply being trained and armed constitutes a sufficient justification and that the donning of a uniform and the carrying of arms signifies a change in moral status of the concerned person and a shift in their status from non-combatant to combatant.[7] Others invoke the boxing ring analogy. This analogy holds that, while punching a man on a street in a civilised society is unfair and immoral, fighting those who voluntarily step into the ring is perfectly justifiable. These 'voluntarists' imply that soldiers renounce their immunity and become legitimate targets once they set foot on the battlefield.[8, 9] Irrespective of the underlying rationale, the task of discrimination is made much more onerous because of the character of modern warfare and of the role which civilians play in aiding technological warfare and the wider 'war machine'.[10]

However, the point here is that no matter how difficult the distinction between combatant and non-combatant is to make in practice, a lack of identifiability does not give a government or its agents the right to kill indiscriminately. The onus is very much placed on the active aggressor to identify combatants and take 'due care' in the process.[11] The implication, then, is that, if there is a particular degree of uncertainty, an attack should not go ahead. If it does go ahead and civilians are killed, responsibility for the attack must generally rest upon the shoulders of those who made the decision. This assumes that we can identify participants in the targeting decision. The same holds true when it comes to tempering the extent and violence of warfare and minimising general destruction. While the discrimination requirement discussed above is concerned with who or what is a legitimate target of war, the proportionality requirement deals more explicitly with how much and what sort of force it is morally permissible to use in seeking to destroy those targets. Even in a battle only involving military targets (a rare occurrence in modern times), it is possible to employ disproportionate force, since one is morally sanctioned to use only the amount of force necessary to achieve the objective at hand.[12] Of course, many believe that unmanned systems allow their users to avoid some of war's suffering by employing highly selective killing. If a small number of legitimate targets can be killed as part of a military operation to avoid further bloodshed, then this makes a great deal of sense according to proportionality. If, however, it brings civilians and their infrastructure into the line of danger as argued in earlier chapters, then it may not be so wise to employ such practices. It will only complicate targeting processes and responsibility ascription.

In any case, the principles of discrimination and proportionality play an important role in aiming to restrict war's violence and range, but there is no obvious reason why an *individual person* ought to take the blame where collective violations of these principles have occurred, as per the Royakker and van Est interpretation. Surely they mean to say that these principles underwrite the requirement that *agents of war*, whether human, machine or some combination thereof, should be held responsible for their actions. Of the many reasons for this, three stand out. The first, and probably the most important, is that states and their agents need to learn from their mistakes and those of others in order to ensure that future wars achieve better results and are conducted more justly.[13] The second claims that the majority of people (especially the victims, their families and the wider affected civilian and/or military community) will not find it satisfactory if, when a casualty occurs, there is nobody or nothing held responsible, i.e. nothing towards which the victim's family etc. can direct attitudes such as blame and resentment or compensation requests. For soldiers, organisations and the like, there is also a respect in which taking responsibility for the role they played in certain actions can help them transition back into the civilian realm with a good conscience or ethical reputation. After all, not all modern soldiers are permanent enlisted members of their respective defence forces that serve overseas, nor are all organisations permanent parts of the military-industrial complex. Therefore, even in peacetime, the right conception of responsibility acts as an inhibitor of immoral conduct in re-asserting the importance of adhering to the laws/customs of war and reminding agents of war and those supporting them that war will eventually end and that they will want to return to civilian life without being plagued by problems that can arise from improper actions and conduct.

That said, when such actions occur, allocating responsibility is just as difficult as differentiating between combatant and non-combatant or making proportionality judgments. As Walzer himself admits, 'it is no longer easy to impose responsibility', especially with 'certain sorts of state action, secretly prepared or suddenly launched'.[14] Thus, it should come as no surprise that many significant war crimes have gone without the assignment of responsibility that is necessary to accord with Walzer's just war theory. In some cases, allocating responsibility is so difficult that Walzer tries to avoid holding specific actors responsible despite his claims about the necessity of doing so. For example, he talks about the many Americans who 'were morally complicitous in our Vietnam aggression', though he says shortly thereafter that he is not 'interested in pointing at particular people or certain that he can do so'.[15] Instead, he only wants to insist that 'there are responsible people … [and that] the moral accounting is difficult and imprecise'.[16] As will be demonstrated in the next section, this moral accounting is only made more difficult in the technological age. Royakker and van Est suggest that this might not be such a big problem, though. They say that 'we do not normally say that whether or not someone may be held responsible after the fact … is a condition of permissibility, at least in the absence of a rule to the contrary'.[17] In other words, they suggest that the permissibility of war is not tied to post-facto responsibility identification unless there is some rule establishing the link. However, war is unlike their purportedly normal circumstances. They ignore the fact that there is a rule of responsibility implicit in just war theory and that, even if there were no rule, this would be a contributor to things like the seemingly endless conflicts discussed in the previous chapter and thus an important reason to develop such a rule and investigate the challenges in adhering to it.

Challenges to Responsibility Attribution in Technologically Enabled Warfare

Moral responsibility in war is about actions, omissions and their consequences. When we read stories in military ethics readers, those worthy of blame include agents failing to adhere to just war principles or to otherwise do 'right thing' by platoon leaders, government or country.[18] It is also about the conditions under which they did the right or wrong thing. To be held responsible in accord with Fischer and Ravizza's[19] landmark account – which is based on the idea of guidance control and that the mechanism that issues the relevant behaviour must be the agent's own and be responsive to reasons – actors must not be 'deceived or ignorant'[20] about what they are doing and ought to have control over their behaviour in a 'suitable sense'.[21] Put more specifically, this means that an agent should only be considered morally responsible if they intentionally make a free and informed causal contribution to the act in question, meaning that they must be aware of the relevant facts and consequences of their actions, have arrived at the decision to act independently and were able to take alternative actions based on their knowledge of the facts.[22] If these conditions are met, we can usually establish a link between the responsible subject and person or object affected, either retrospectively or prospectively (the latter will be the focus of the final section). However, technologically enabled warfare of the unmanned type presents various challenges for these standard accounts of moral responsibility. For the sake of a complete exposition and refutation of Sparrow's claim that the responsibility gap presents an insurmountable threat, it is necessary to take a closer look at how semi-autonomous military technologies, generally defined, can complicate responsibility attribution in warfare.

There are many barriers to responsibility attribution in the military domain and many are so closely interrelated that it makes providing a clear and lucid discussion quite problematic. The most important for present purposes is associated with the subject's causal contribution to the action in question. According to the above referred-to account, for an agent to be held responsible, they must have exerted due influence on the resulting event. What is 'due' will be subject to further reflection in the remaining sections, but there is little to be gained from blaming someone or something for an unfortunate event about which they/it legitimately had no other choice or over which they/it had no control. That acknowledged, the employment of modern warfighting technologies based on complex computing and information technologies can lead us to lose our grasp of who is responsible, because it obscures the causal connections between an agent's actions and the eventual consequences. When utilising complex technologies, tracing the sequence of events that led to a particular event usually leads in a great number of directions.[23] The great majority of technological mishaps are the product of multifaceted mistakes commonly involving a wide range of persons, not limited to end users, engineers and technicians. For those looking from the outside in, it can be very difficult – and some (like Sparrow) might say impossible – to identify contributing agents. This difficulty in identifying contributing agents is Dennis Thompson's so-called 'problem of many hands'.[24] This problem should not be confused with the 'responsibility gap' that will soon be addressed, because it is not as deflationary and falls short of the complete abdication of responsibility.

Added to the problem of many hands is the physical distance that warfighting technologies often create between agents and the consequences or outcomes of their actions. This further blurs the causal connection between action and event. Batya Friedman earlier noted this effect in an educational setting which encourages young people to become responsible members of the electronic information community.[25] The challenge has been reinvigorated with the development and deployment of unmanned systems in the military setting and the employment of distanced drone operators. It is these war-making agents that now need to be encouraged to play a responsible role in network-centric operating environments. As argued more extensively in chapters 5 and 6, unmanned systems technologies – more than any other material technology – extend the reach of military activity through both time and space. While enabling a state's military force to defend itself over a greater range than they would otherwise be able to may be morally permissible, or even praiseworthy, and may be in line with the social contract, this remoteness can also act to disassociate them from the harm that they cause. As noted earlier, it has long been understood that there is a positive relationship between the physical and emotional distance facilitated by technological artefacts and the subsequent ease of killing.[26] When someone uses an unmanned aircraft operated from a control station on the ground in the US to conduct military operations in the Middle East, the operator might not be fully aware of how the system and its munitions will affect the local people and may not experience or fully appreciate the true consequences of their actions.[27] This has a direct bearing on their comprehension of the significance of their actions and has a mediating role when it comes to considering the extent to which they are responsible.

This mediation of responsibility has much to do with the fact that unmanned systems and the sensors that they carry can actively shape how moral agents perceive and experience the world at large, which further impacts upon the conditions for imposing moral responsibility. In order to make the appropriate decisions which are sanctioned by just war theory, a moral agent must be capable of fully considering and deliberating about the consequences of their actions, understanding the relevant risks and benefits they will have and to whom they will apply. This, in turn, calls for them to have adequate knowledge of the relevant facts. While McMahan and others have offered accounts,[28] it remains unclear what epistemic thresholds ought to apply here, but what is generally accepted is that it is unfair to hold someone responsible for something they could not have known about or reasonably anticipated. The capability of unmanned systems and other intelligence gathering technologies is importantly relevant, because in some respects they assist the relevant users in deliberating on the appropriate course of action by helping them capture, collate and analyse information and data.[29] In their sales demonstrations to the military, for example, representatives of the drone industry typically argue that their piece of military hardware will grant them the opportunity to see 'beyond the next hill' in the field and 'around the next block' in congested urban environments, enabling them to acquire information that they would not otherwise have access to without incurring significantly greater risk.[30] This may well be true with respect to some systems, and these would allow operators greater reflection on the consequences of their tactical decisions. However, with the technical, geographical and operational limits discussed in Chapter 5, there are many respects in which these systems preclude one from gaining a view of the 'bigger picture' and may alter an operator's resulting action/s, perhaps limiting responsibility.

Many intelligent military systems have such complex processes that they get in the way of assessing the validity and relevance of the information they produce or help assess and, as such, they can actually prevent a user from making the appropriate decision within an operational context and therefore have a direct impact on their level of responsibility. A consequence of this complexity is that people have the aforementioned tendency to rely either too much or not enough on automated systems like those we increasingly find embedded into unmanned aircraft or their control systems, especially when in the time-critical and dynamic situations which are characteristic of modern warfare.[31] The USS Vincennes most shockingly illustrated this during its deployment to the Persian Gulf amid a gun battle with Iranian small boats. Although this warship was armed with an Aegis Combat System, which is arguably one of the most complex and automated naval weapons system of its time (it can automatically track and target incoming projectiles and enemy aircraft), the USS Vincennes misidentified an Iranian airliner as an F-14 fighter jet and fired upon it, killing nearly 300 people.[32] Post-accident reporting and analysis discovered that overconfidence in the abilities of the system, coupled with a poor human-machine interface, prevented those aboard the ship from intervening to avoid the tragedy. Despite the fact that disconfirming evidence was available from nearby vessels as to the nature of the aircraft, it was still mischaracterised as a hostile fighter descending and approaching them at great speed. In the resulting investigation, a junior officer remarked that 'we called her Robocruiser … she always seemed to have a picture and … [always] seemed to be telling people to get on or off the link as though her picture was better'.[33] The officer's impression was that the semi-autonomous system provided reliable information that was otherwise unobtainable. In this case, at least, such a belief was incorrect. The system had not provided otherwise unobtainable information, but rather misleading information. It is therefore questionable whether the war-making agent has a more comprehensive understanding of the relevant state of affairs because of the employment of advanced military technology or whether her/his understanding and knowledge are less accurate.[34] That is, it is unclear whether the attribution of moral responsibility is enhanced or threatened. The view advanced here is that, even though there may be an aggregate increase in the amount of information that is accessible, there is a morally relevant decrease in understanding which single piece of information ought to influence autonomy of action and the resulting decision making, even when the bulk of information is clear and accurate. The implication is that operators of sophisticated systems might be held to high standards of responsibility on the basis that they had access to a great deal of relevant information when, in fact, the provision of this information may have clouded their judgement, meaning that they are less responsible.

It must also be added that advanced technologies may exert a certain level of influence over their users in a way that might be unclear or even immeasurable.[35] This sort of control is not implicit in the technology itself, but rather exerted through the design process and the way in which alternative moral options are presented for human action. Semi-autonomous military technologies help to centralise and increase control over multiple operations, reducing costs and supposedly increasing efficiency. However, there is a limit to how much control a human being can exert and, in reality, this 'increased control' can only be achieved by outsourcing some low-level decisions to computerised processes and leaving the human to make a choice from a more limited range of actions. In other words, some military technologies are designed with the explicit aim of making humans behave in certain ways, further mediating the imposition of responsibility. However, note that we are still a long way from saying that we cannot attribute responsibility in such cases.

The Alleged Responsibility Gap in Automated Warfare

In the previous section, we saw how developments in military technology have led to a partial loss of influence on the part of operators or users and hold broader implications for the attribution of moral responsibility more generally, namely by limiting the operator's responsibility and perhaps causing us to consider the redistribution of the remaining share of responsibility. In the following section of this chapter, we shall see how many of the problems described above are only being exacerbated as unmanned systems become more computerised. These problems and others come together in fully automated warfare to create a problem that Sparrow and others see as significantly more serious than any of those already discussed, posing what is supposedly an insurmountable threat to the responsibility framework embedded within just war theory. As a final step towards fully understanding and refuting the nature and implications of this problem, it will be necessary to discuss the arguments drawn on by Sparrow, namely those originally put forward by Andreas Matthias.[36]

Matthias argues that the further we progress along the autonomy continuum, the closer we are to undermining society's centuries-old effort to establish rule systems in order to attribute responsibility.[37] He says that with non-autonomous systems, it is relatively safe to take the use of a machine to signal that the operator has read the user manual and assumes responsibility for its use, except in cases where the machine fails to operate with the predefined limits.[38] Thus, the user has control and is responsible for the actions and events that come from the normal operation of the system, but if it explodes or does something which was not stated in the manual when it should have been, we ought to blame the manufacturer. We know from the preceding discussion that, when machines capable of being ever so slightly autonomous are introduced, because of the rigidity and limited nature of non-autonomous systems, moral responsibility is complicated. Indeed, the agent responsible for operating this machine loses an element of control over the system. What happens if we progress further? Matthias argues that, if a NASA technician was operating a semi-autonomous space vehicle and the vehicle falls into a crater between inputs because of long response times, we should not consider the technician responsible.[39] Task-autonomous unmanned systems create a buffer between agent and system as well as another buffer between action and event, giving their operators the potential ability to increase their workload by operating multiple drones, but at the cost of understanding and situational awareness. The locus of responsibility categorically shifts away from the operator. However, the real problems arise when it comes to intelligent machines which are capable of adapting and learning new skills. That is, robotic systems for which unpredictability in their operations is a feature rather than a computer glitch or technical failure.[40] Matthias asks us to imagine how we would impose moral responsibility if we were to revisit the space vehicle case and stipulate that it will not be remotely controlled from earth, but rather have its own integrated navigation and control system, capable of storing data in its internal memory, forming representations and taking action from these.[41] It should, therefore, be able to record video imagery and estimate the difficulty of crossing any familiar terrain. He asks, in this revised case, whom we should hold responsible if the vehicle were to once again fall into the crater?

Sparrow subsequently takes up this question in the context of his discussion of unmanned systems.[42] He has us imagine that a drone, directed by sophisticated artificial intelligence, bombs a platoon of enemy soldiers who have indicated their intent to surrender. Who should we hold morally responsible for a particular event when the decision to bomb is made by an autonomous weapons system without a human operator? The reader's first intuition is probably to say that the responsibility for any moral crimes or just war violations rests with the developer of the weapon. However, Sparrow objects to this by relying on the user manual analogy. This analogy says that this would be unfair if it is a declared system limitation that the machine may attack the wrong targets in some percentage of cases.[43] If this is the case, he suggests it may be the responsibility of the user (since s/he is assumed to have read the manual). Secondly, he says that to hold the programmers or manufacturer responsible for the actions of their creation, once it is turned on and made autonomous, would be analogous to 'holding parents responsible for the action of their children once they have left their care'.[44] Sparrow assumes this is wrong and that it naturally leads us to consider holding the commanding officer responsible. Yet again, he views this as unfair and thinks that to do so calls into question the nature of our 'smart' weapons. If the machines start to make their own targeting decisions, he suggests that there will come a point at which we cannot hold the commanding officer responsible for any of the ensuing deaths.[45] It will be argued in the next section that Sparrow is mistaken about the assumed wrongness of sharing responsibility and that we are not yet at the tipping point he describes.

The final possible locus of responsibility under Sparrow's account, however, is the machine itself. Moral responsibility is typically attributed to moral agents and, at least in the Anglo-American philosophical tradition, moral agency has been reserved for human beings. The reason being that unlike the majority of animals, rational human beings are seen as able to freely deliberate about the consequences of their actions and choose to act in one way or other, meaning that they are originators of *morally significant* actions. Although some people tend to anthropomorphise military robots and notable philosophers like Dennett and Sullins have argued that they could be classed as moral agents,[46] Sparrow argues that they should not and objects to the idea that they could have the kind of capacities that make human beings moral agents. He argues that it is unlikely that they will ever have the mental states, common sense, emotion or expressivity equivalent to those of humans, and that, if they do develop these things, it would undermine the whole point of utilising robots instead of human beings.[47] According to the argument, they would hold a moral status equivalent to that of human beings. But this is disputable, as it could be argued that, while artificial moral agents may be worthy of moral consideration, they would still hold different status to biological moral agents by virtue of some natural/artificial distinction that gives greater weight to means of creation. Even if robots do not acquire human-level moral status, lower levels of machine autonomy may be sufficient for us to hold robots responsible. However, Sparrow holds that no robot can be held responsible because they cannot suffer.[48] This presupposes that suffering is a requirement for responsibility, a presupposition not supported by the responsibility framework embedded in just war theory, but we will return to this point after having summarised the alleged problem.

For Sparrow, like Matthias before him, we have reached or are about to reach an important impasse. We already have many machines in development and a limited number of those, which are in use, are task-autonomous and can decide on a course of action in some limited scenarios without any human input. Going forward, all indications point to there being machines with rules for action which are not fixed by their manufacturers during the production process and which are open to be changed by the machine itself during its operation. That is, these machines will be capable of learning from their surroundings and experiences. Conventionally, there are several loci of responsibility for the actions of a machine, but both Matthias and Sparrow argue that these robots will bring about a class of actions for which nobody is responsible, because no individual or group has sufficient control of these systems.[49] These cases constitute Matthias's 'responsibility gap'. At first blush, it might seem that this is basically the problem of many hands – the classical problem described earlier but with new relevance to the emergence of unmanned systems and the prospect of fully autonomous weapons systems. But to assume this would be mistaken. The argument advanced by Matthias and Sparrow is not that we cannot identify who is responsible, but simply that nobody is responsible. Sparrow would likely argue that, if there is any problem identifying the relevant persons, it is because they do not exist. In his article on corporate responsibility, Philip Petitt refers to the matter described as the 'problem of no hands'.[50] This is a slight but important twist on Thompson's more familiar 'problem of many hands', as described earlier and characterised by the widespread relinquishment of moral responsibility.

The problem that proponents of the responsibility gap put forward is fairly straightforward, though certainly not indisputable. However, to differentiate his argument from Matthias's, Sparrow suggests that we might better conceptualise his dilemma if we consider another case in warfare where the waters are somewhat muddied: the use of child soldiers.[51] This analogy is outlined because it will be useful in problematising Sparrow's argument, though the similarity between child and machine learning is not as great as Sparrow indicates. He says that like robots, one of the many reasons why it is unethical to utilise children in combat is that it places decisions about the use of force in the hands of agents that cannot be held responsible for them. According to him, child soldiers lack full moral autonomy, but they are clearly autonomous in some respect and 'certainly much more autonomous than any existing robot'.[52] He goes on to say that, while they are not capable of understanding the full moral dimensions of what they do, they possess sufficient autonomy to ensure that those who order them into action do not or cannot control them, presenting problems for any effort to hold those who gives the orders exclusively personally responsible for the child soldiers' actions.[53] The idea Sparrow advances is that there is a conceptual space in which child soldiers and military robots are sufficiently autonomous to make the full attribution of responsibility to an adult or conventional moral agent problematic, but not autonomous enough to be held fully responsible themselves. Sparrow argues that his opponents try to close this space by stipulating that the relevant entities hold more or less responsibility than they should and thus fit within one of the polar boundaries, but that this does not adequately or fairly resolve the problem. He thinks that we should, in fact, ban the use of autonomous weapons altogether.[54] The next section will propose that we can actually handle this problem by moving to a more collective, pragmatic and forward-looking notion of shared responsibility.

Toward a Revised Notion of Responsibility

Having outlined the importance of responsibility in the just war framework, explored some of the challenges that semi-autonomous systems pose for responsibility attribution and described the dilemma over responsibility for autonomous unmanned systems, the need for a revised notion of responsibility should be clear. But to clarify, the arguments of both Matthias and Sparrow hinge on three basic premises. The first is that programmers, manufacturers, commanding officers and the like may not be able to foresee what an autonomous robot, capable of learning, will do in the highly complex and dynamic military operating environment. The second is that either independently of or related to the fact that none of these agents are able to exert full control over the development or subsequent deployment of these systems, harm to others may eventually occur. The third is that an agent can only be held responsible for these harms if they have control in the sense that they have an awareness of the facts surrounding the action that leads to the harm and are able to freely manipulate the relevant causal chains based on these facts. The conclusion stipulates that since this is not the case as it pertains to programmers, manufacturers or commanding officers, there is some sort of moral void created by the deployment of these systems, one that cannot be bridged by our traditional concepts of responsibility. While the problem is clear, it is not obvious that the overall conclusion can be accepted at face value or that any individual premise is correct. There are a number of points at which the alleged responsibility gap can be overcome, or, at least, a number of premises that can be called into question in order to cast doubt over the supposed insurmountability of the problem at hand.

In discussing the nature of this alleged responsibility gap and showing its inadequacy as a justification for a moratorium on autonomous systems, it is important to point out that the scope of the conditions for imposing responsibility have been overstretched or considered in too wide a frame. As will soon be shown with reference to the idea of shared responsibility, it is not impossible to impose responsibility in situations in which no individual has total control over the function of an autonomous system. Both Matthias and Sparrow go too far in suggesting that programmers, manufacturers or commanding officers are freed from any form of responsibility because they do not have total control over the operation or manufacture of unmanned systems. An appeal to common sense should reveal that it is absurd and potentially dangerous to identify the failings of multiple individuals in their development and deployment and then deny their moral responsibility.[55] To do so is deny an opportunity for the rectification of past, present and future wrongs. As opposed to what might be stipulated by strict liability law, such a strong sense of control is not necessary for the imposition of some degree of moral responsibility. The relevant programmers, designers, manufacturers and commanding officers are all responsible to some degree or extent. Take Sparrow's claim that to hold the programmer of a dynamic learning machine responsible for its actions would be analogous to holding a parent responsible for the actions of their child once they are out of their care. While there are limits to any such analogy because of the varied learning mechanisms employed by child and machine, he seems to ignore the fact that parents are at least partially responsible for preparing their children for that moment when they leave their care and become independent. In much the same way, the developers of unmanned systems hold significant responsibility for ensuring that their robots can operate as desired once given independence, something that is still a long way off in the majority of cases. Also, take the case of the commanding officer and continue with the parenting analogy. When a parent is teaching their child how to drive, for instance, the parent places the child in an area where s/he can learn the necessary skills without risking her/his own safety or that of anyone else. Similarly, the commanding officer of an autonomous system has a responsibility to ensure that the system has been thoroughly proto-type tested or is placed in an appropriate learning or test bed environment until such time that it performs at least as well as a manned system or until the chance of any serious harm occurring is so tiny that we can deal with it via the 'functional morality' described later on, which recognises that engineers and manufacturers will often choose to release intelligent machines with remnant unpredictability and that reprogramming for minor errors can assume the place of punishment.

The machine's path to full autonomy is a long (if not impossible) one and Sparrow points to this using the child warrior analogy. Machines will not just 'wake up' as is depicted in films about human-hating 'terminators'. Indeed, there is simply no way in which someone could deliberately create such an entity without a collective effort on the scale of the Manhattan project. The general lesson to be drawn from this is that all the involved agents and any others associated with the use of unmanned systems (including the user in the case of semi-autonomous systems) retain a share of responsibility, even though they may claim that they were not in complete or absolute control. It would be foolhardy, or even dangerous, to conclude from the observation that responsibility is obscured by the use of unmanned weaponry that nobody is, or ought to have been, held to account and that it is impossible to deal with the case of autonomous systems. On the contrary (and as others have argued in relation to informatics more generally)[56] we are at such an important junction in the development of unmanned systems that we have reason to adjust and refine our conception of moral responsibility and to leave behind the idea that the imposition of moral responsibility relies on agents having full control over all aspects of the design or deployment of advanced robotics. This is because these systems are so complex that few of the design, development or deployment related decisions are made on an individual basis. Why concentrate on the intentions and actions of humans alone in our moral evaluation of these systems when no human exerts full control over the relevant outcomes? We need to move away from the largely insufficient notion of individual responsibility, upon which we typically rely, and move towards a more complex notion of collective responsibility, which has the means and scope to include non-human action. That is, it must be a holistic approach which is capable of acknowledging the contribution of various human agents, systems and organisations or institutions. The need to update our moral values and associated notions of responsibility will become more important as the technology develops, the risks increase and the number of potential responsibility-related problems accumulate.

It is worth noting that others, foreseeing the difficulties that we are now facing with the development of things like intelligent autonomous robots, have already thought about calls for change in the way we think of responsibility. For instance, both Daniel Dennett and Peter Strawson have long held that we should conceive of moral responsibility as less of an individual duty and more of a role that is actively defined by pragmatic group norms.[57] This argument in endorsed here, primarily because more classical accounts raise endless questions concerning free will and intentionality that cannot be easily resolved (if at all) from a practical perspective aimed at achieving results here and now. This sort of practical account has the benefit of allowing non-human entities, such as complex socio-technical systems and the corporations that manufacture them to be answerable for the harms to which they often cause or contribute. It seems to require that we think in terms of a continuum of agency between non-moral and full moral agents, with the sort of robots we are concerned with here falling just short of the latter. This pragmatic (or functional) approach also allows for the fact that agency develops over time and shifts the focus to the future appropriate behaviour of complex systems, with moral responsibility being more a matter of rational and socially efficient policy that is largely outcomes-focused. For our purposes here, it is useful to view moral responsibility and this pragmatic account in line with the social contract argument put forward much earlier. That is, we should view moral responsibility as a mechanism used by society to defend public spaces and maintain a state of relative harmony, generated under the contract by the power transferred from the individuals to the state. The end of this responsibility mechanism is, therefore, to prevent any further injury being done to society and to prevent others from committing similar offences. While it might be useful to punish violators of the social contract and the just war theory which it has argued follows from the social contract, it is not strictly necessary, nor is it necessarily the best approach to preventing harm in all scenarios. As Jeroen van den Hoven and Gert-Jan Lokhorst have argued, treatment is in many cases an equally effective option for the prevention of harm and one that we can apply to non-human agents in different forms, whether it is psychological counselling in humans or reengineering or reprogramming in the case of robots.[58] This is important because it means that there is sufficient conceptual room between 'operational morality and genuine moral agency'[59] to hold responsible or, in Floridi's language, hold morally accountable, artificial agents which are able to perform some task and assess its outcome.[60]

Scholars in the drone debate also seem to have become fixated on a backward-looking (retrospective) sense of responsibility, perhaps because even engineers and programmers have tended to adopt a malpractice model focused on the allocation of blame for harmful incidents. However, an effective and efficient responsibility mechanism that remedies the supposed gap should not only be about holding someone responsible only when something goes wrong. Therefore, this backward-looking sense of responsibility must be differentiated from forward-looking (prospective) responsibility, a notion that focuses more on capacity to effect change than blameworthiness or similar.[61] That is, at some point, we must stop thinking purely about past failures to take proper care and think about the reciprocal responsibility to take due care in future action. This is because in debates about real world problems such as the deployment of increasingly autonomous unmanned weaponry, we will also want our conception of responsibility to deal with *potential* problems. To this end, we can impose forward-looking or prospective responsibility to perform actions from now on, primarily in order to prevent undesirable consequences or to ensure a particular state of affairs obtains more effectively and efficiently than through the alternative means of backward-looking models. It establishes a general obligation on the part of all those involved in the design, manufacture and use of unmanned systems to give regard to future harms. Admittedly, as Seamus Miller has pointed out in the case of computing, it is difficult to reach any solid conclusion on how far into the future they are required to foresee.[62] That said, two things are clear in the case of military robotics: first, if the recent troubles plaguing unmanned systems are any indicator, agents have reasonable grounds to expect the unexpected. Second, once fully autonomous systems are developed and deployed, no amount of policy making will stop their spread.

For the latter reason in particular, we have to think more carefully about where the majority of the forward-looking responsibility falls. In discussing the claim that wealthy countries must do more than comparatively poor countries to combat climate change and making use of Kant, James Garvey argues that in much the same way that 'ought implies can', 'can implies ought' in a range of other circumstances where the financial or political means behind the 'can' have contributed to the problem that ought to be corrected or mitigated.[63] This seems also to hold true in the responsibility debate with which we are engaged. Generally speaking, the more power an agent has and the greater the resources at their disposal – whether intellectual, economic or otherwise – the more obliged s/he is to take reasonable action when problems arise.[64] Jessica Fahlquist has proposed a specific approach to identifying the extent of a person's obligation in relation to environmental protection based on varied levels of capacity to contribute to social causes, and given that power to enact change varies within the military and military-industrial complex to much the same extent as in the environmental world (and with some convergence), there is little reason why this should not be applied to the drone debate and extended to cover the manufacturers of unmanned systems as well as the governments that regulate them.[65] The companies of the military-industrial complex are in a unique position, and have it well within in their power, to anticipate risks of harm and injury and theorise about the possible consequences of developing learning systems. The costs of doing so after the fact are great and many. Moreover, manufacturers are best positioned create opportunities for engineers to do what is right without fear of reprimand, whether that would be going ahead as planned, designing in certain limitations in the system or simply refusing to undertake certain projects. It therefore seems reasonable to impose forward-looking responsibility upon them. However, we know that profit can sometimes trump morality for these collective agents of the military-industrial complex, so concerned parties should also seek to share forward-looking responsibility and ascribe some degree of responsibility to the governments which oversee these manufacturers and set the regulatory framework for the development and deployment of their product, should manufacturers fail to self-regulate and establish appropriate industry standards.

Note that, while it may seem that the point advanced in this chapter is pitched against Sparrow's argument for a prohibition on the development and subsequent use of these systems, this is only partly true. The objection here is only to the antecedent of his claim or the vehicle that he uses to reach his final conclusion against the use of 'killer drones'. That is, it is not at all obvious that Sparrow needs to make the very bold claim that nobody can be held responsible for the use of military robots that fall between having operational autonomy and genuine moral autonomy. It seems that many other agents are sufficiently responsible and that, through embracing an instrumental approach toward backward-looking responsibility and combining it with a forward-looking account of responsibility put forward earlier, it is possible to distribute responsibility fairly.[66] It may also be that some of the relevant agents, namely governments, reach the same conclusion as Sparrow. That is, while states and their militaries have a contractual obligation to effectively and efficiently protect the citizens who grant them power, it may turn out that this would mean avoiding the use of unmanned systems in some circumstances and, indeed, there is some evidence of this in the implications and consequences described in earlier chapters. In fact, as unlikely as it seems, time may prove that unmanned systems pose such a problem that it warrants making a concerted effort to form an international treaty or a new Geneva Convention banning their use in all but very particular cases, perhaps where there is clearly demarcated 'kill zone' in some remote part of the world. However, this is not yet clear and in light of the features of the revised account of responsibility advocated here, there is no intrinsic responsibility gap that warrants a prohibition of the use of unmanned systems, at least not without reference to the problems discussed in other chapters.

Critics are likely to argue that any account which allocates too much responsibility to organisational or governmental actors will erode the sense of personal responsibility that individuals feel and will not thus have the desired effect of improving the attainment of just outcomes with unmanned systems. What will happen, they may ask, if individual parties to collective atrocities are excused from the responsibilities that they would otherwise have under traditional notions of responsibility? There are a few ways to respond to this worry. The first consists in stressing that individuals are not in fact freed or excused from responsibility according to either the backward-looking or the forward-looking model. Individual agents will still be held causally responsible for their part in any war crimes under the functional/pragmatic account, so if a programmer or a commanding officer gets an entire village blown up by mistake, they will be held accountable. They will also be encouraged to ensure that systems are designed, developed, tested and used in the desired way through the imposition of forward-looking responsibility. The second way to respond is to reaffirm that, because individual agents are the core units of the social contract, it would obviously be ideal if they were to gradually begin to embrace the right values and do the right thing, but allow that, in some circumstances, it is fairer or more effective to distribute the burden of action between individuals, institutions and governments. This is due to the fact that, in the short to medium term, it must be recognised that both human and non-human agents will make mistakes that will lead to violations of some of the principles of the just war theory. As a consequence, the greatest share of responsibility must be ascribed to the most capable agents in the relevant scenario. In trying to highlight the oversimplified analysis of duties in rights-based theories, Henry Shue suggests that in some circumstances, we must look beyond individuals and distribute responsibility to institutions in the most effective and efficient fashion relative to the time we have as well as the nature and severity of the problem.[67] For it is indeed institutions, such as judicial systems and polices forces, upon which the duty to provide physical security ultimately falls. In the case of Sparrow's 'killer robots', it seems best to impose much of the forward-looking responsibility to the government or its relevant standards departments, just because the government is best placed to ensure that systems are designed to rule out or reduce the impact of such mistakes or that measures are put in place to do the same thing. Again, it must be stressed that this does not mean that individual agents or corporations with the relevant capabilities and resources are excused from efforts to achieve the desired effects or their share of responsibility, but merely that governmental agents ought to take greater efforts because the proposed account of responsibility can track their role and they can be held responsible for some violations of the just war theory.

Conclusion

This chapter began by exploring the need for a clear account of responsibility in just war theory and countering the claim that there is some sort of explicit requirement to hold a single individual responsible. It indicated that there is, less specifically, an implicit responsibility component which stipulates that agents of war – whether human, non-human or some combination thereof – must be held responsible for violations of the just war theory no matter how difficult the moral accounting. In the second section, it was demonstrated that technology generates a number of barriers to attribution of responsibility, from distancing users from their sense of responsibility to obscuring causal chains, making it more difficult to identify where a moral fault lies. In the third, it was outlined how these issues and others come together in the case of fully automated unmanned warfare to create what Sparrow – following Matthias – alleges is a 'responsibility gap', or a class of actions for which nobody/nothing is supposedly responsible. The final section laid the foundations for a theory of responsibility, which revolves around the idea that action and responsibility can be distributed amongst human and non-human agents or some combination thereof. More work is needed to reveal exactly what this new theory of responsibility will look like and to determine its precise implications, but if nothing else, this chapter has hopefully demonstrated that, while ascribing responsibility in the case of autonomous systems is more complex and troubling than in the case of semi- and non-autonomous systems, it is by no means an insurmountable problem.

Notes

1 Walzer, *Just and Unjust Wars: A Moral Argument with Historical Illustrations*, 287–8.

2 Many of these theorists belong to the International Committee for Robot Arms Control (ICRAC), which has issued a call to ban autonomous lethal robots in which the decision to apply violent force is made autonomously. The ICRAC starts its justification for a ban by stating 'we are concerned about the potential of robots to undermine human responsibility in decisions to use force, and to obscure accountability of the consequences'. See International Committee for Robot Arms Control, 'Scientists's Call', http://icrac.net/

call/; also see Bill Keller, 'Smart Drones', *The New York Times* (2013), http://www.nytimes.com/2013/03/17/opinion/sunday/keller-smart-drones.html?pagewanted=all.

3 Sparrow, 'Killer Robots', 67.

4 Ibid.

5 Royakkers and van Est, 'The Cubicle Warrior: The Marionette of Digitalized Warfare', 290.

6 J. Fieser and B. Dowden, 'Just War Theory', *The Internet Encyclopedia of Philosophy* (2007), http://www.iep.utm.edu/j/justwar.htm.

7 Jeff McMahan, 'The Ethics of Killing in War', *Ethics* 114, no. 1 (2004): 55.

8 Ibid., 56.

9 Both of these accounts are troubled by the problems associated with delineating the boundaries of rights and their respective bearers.

10 For a relevant discussion, see Walzer, *Just and Unjust Wars: A Moral Argument with Historical Illustrations*, 146.

11 Ibid., 156.

12 Ibid., 129.

13 Royakkers and van Est, 'The Cubicle Warrior: The Marionette of Digitalized Warfare'.

14 Walzer, *Just and Unjust Wars: A Moral Argument with Historical Illustrations*, 301.

15 Ibid., 302–3.

16 Ibid.

17 Royakkers and van Est, 'The Cubicle Warrior: The Marionette of Digitalized Warfare', 291.

18 For many such examples, see Coleman, *Military Ethics: An Introduction with Case Studies*.

19 John Martin Fischer and Mark Ravizza, *Responsibility and Control: A Theory of Moral Responsibility* (Cambridge: Cambridge University Press, 1998).

20 Ibid., 13.

21 Ibid., 12–13.

22 While the most common factors have been noted here, as found in Katinka Waelbers, 'Technological Delegation: Responsibility for the Unintended', *Science and Engineering Ethics* 15, no. 1 (2009): 58–60. There is some disagreement concerning the conditions required to hold someone morally responsible. For further discussion, see the numerous perspectives put forward in the following: Hans Jonas, *The Imperative of Responsibility. In Search of an Ethics for the Technological Age* (Chicago: University of Chicago Press, 1984); John Martin Fischer and Mark Ravizza, eds, *Perspectives on Moral Responsibility* (Ithaca: Cornell University Press, 1993).

23 Merel Noorman, 'Computing and Moral Responsibility', http://plato.stanford.edu/archives/fall2012/entries/computing-responsibility/.

24 This term was coined by Dennis Thompson, writing on the topic of the moral responsibilities of political office holders within large bureaucracies. For more, see Dennis Thompson, *Political Ethics and Public Office* (Cambridge: Harvard University Press,

1987); Dennis Thompson, 'Moral Responsibility and Public Officials: The Problem of Many Hands', *American Political Science Review* 74, no. 4 (1980).

25 Batya Friedman, 'Moral Responsibility and Computer Technology', in *Annual Meeting of the American Educational Research Association* (Boston, MA1990).

26 Grossman, *On Killing: The Psychological Cost of Learning to Kill in War and Society*, 97.

27 Waelbers, 'Technological Delegation: Responsibility for the Unintended', 52.

28 McMahan, *Killing in War*. See, in particular, chapters 2 and 3.

29 Shoshana Zuboff, 'Automate/Informate: The Two Faces of Intelligent Technology', *Organizational Dynamics* 14, no. 2 (1985): 8–10.

30 United States Department of Defense, 'FY2009-2034 Unmanned Systems Integrated Roadmap', 2.

31 Mary L. Cummings, 'Automation Bias in Intelligent Time Critical Decision Support Systems', in *AIAA 1st Intelligent Systems Technical Conference* (Chicago, 2004).

32 For a more complete discussion of this incident and the sequence of events leading up to it, see the following: Regan H. Campbell and Geoffrey Eaton, 'It's All About You: Implementing Human-Centered Design', in *Human Performance Enhancement in High-Risk Environments: Insights, Developments, and Future Directions from Military Research*, ed. Paul O'Connor and Joseph Cohn (Santa Barbara: ABC-CLIO, 2011), 214–15; Colin S. Gray, 'AI at War: The Aegis System in Combat', in *Directions and Implications of Advanced Computing*, ed. D. Shculer (New York: Ablex, 1997).

33 W. Rogers and S. Rogers, *Storm Center: The USS Vincennes and Iran Air Flight 655* (Annapolis: Naval Institute Press, 1992), 133.

34 Noemi Manders-Huits, 'Moral Responsibility and IT for Human Enhancement' (paper presented at the Association for Computing Machinery Symposium on Applied Computing, Dijon, 2006).

35 Ibid.

36 Andreas Matthias, 'The Responsibility Gap: Ascribing Responsibility for the Actions of Learning Automata', *Ethics and Information Technology* 6 (2004).

37 Ibid., 175.

38 Ibid., 175–6.

39 Ibid.

40 Jason Millar and Ian Kerr, 'Delegation, Relinquishment and Responsibility: The Prospect of Expert Robots', in *We Robot* (Coral Gables: University of Miami School of Law, 2012), 6.

41 Matthias, 'The Responsibility Gap: Ascribing Responsibility for the Actions of Learning Automata', 176.

42 Sparrow, 'Killer Robots', 69.

43 Ibid., 70.

44 Ibid.

45 Ibid., 70–71.

46 John P. Sullins, 'When Is a Robot a Moral Agent?', *International Review of Information Ethics* 6, no. 1 (2006); Daniel D. Dennett, 'When HAL Kills, Who's to Blame?

Computer Ethics', in *HAL's Legacy: 2001's Computer as a Dream and Reality*, ed. D. G. Stork (Cambridge: MIT Press, 1997).

47 Sparrow, 'Killer Robots', 71–3.

48 Ibid., 72.

49 Ibid., 73; Matthias, 'The Responsibility Gap: Ascribing Responsibility for the Actions of Learning Automata', 177.

50 Philip Petitt, 'Responsibility Incorporated', *Ethics* 117, no. 2 (2007): 198.

51 Sparrow, 'Killer Robots', 73–4.

52 Ibid., 73.

53 Ibid., 74.

54 Ibid.

55 Donald Gotterbarn, 'Informatics and Professional Responsibility', *Science and Engineering Ethics* 7, no. 2 (2001): 225.

56 Manders-Huits, 'Moral Responsibility and IT for Human Enhancement'.

57 Daniel D. Dennett, 'Mechanism and Responsibility', in *Essays on Freedom of Action*, ed. T. Honderich (Boston: Routledge and Kegan Paul, 1973); Peter F. Strawson, 'Freedom and Resentment', in *Freedom and Resentment and Other Essays* (Methuen: London, 1974).

58 Lokhorst and van den Hoven, 'Responsibility for Military Robots'.

59 Wallach and Allen, *Moral Machines: Teaching Robots Right from Wrong*, 10.

60 Luciano Floridi and J. Sancers, 'The Foundationalist Debate in Computer Ethics', in *Readings in Cyberethics*, ed. R. Spinello and H. Tavani (Massachusetts: Jones and Bartlett, 2004).

61 Gotterbarn, 'Informatics and Professional Responsibility', 227.

62 Seumas Miller, 'Collective Responsibility and Information and Communication Technology', in *Information Technology and Moral Philosophy*, ed. Jeroen van den Hoven and John Weckert (Cambridge: Cambridge University Press, 2008).

63 James Garvey, *The Ethics of Climate Change: Right and Wrong in a Warning World* (Bloomsbury: New York, 2008), 86.

64 Jessica Nihlén Fahlquist, 'Moral Responsibility for Environmental Problems – Individual or Institutional?', *Journal of Agricultural and Environmental Ethics* 22, no. 2 (2008): 123.

65 Ibid.

66 No exact division of responsibility is advocated here, since this book is not concerned with jurisprudence or specific cases.

67 Henry Shue, 'Mediating Duties', *Ethics* 98, no. 4 (1988): 696–8.

Chapter 10
Conclusion

Despite the diversity of topics covered in the previous chapters, there is a definite attitude and master argument running throughout this book and it is one that expresses the difficulty in balancing the social, political and moral issues of the day with the need for national defence. The narrative trajectory began with a cautionary tale about claims that unmanned systems might make war less savage, impotent or even obsolete (Chapter 1). The same claims, it was said, were made about a range of other military-technical innovations, including firearms, tanks, bombs and missiles. Next was a technological survey and discussion of the progressively autonomous nature of military-technical innovations (Chapter 2). While we are far from the singularity that could give rise to a genuinely autonomous robot with the abilities of a human warfighter, the enabling technologies are developing rapidly and semi-autonomous systems are increasingly becoming intelligent enough to be commercially viable and militarily useful.

Chapter 3 outlined how these technologies will enable many governments and their military forces to deploy force with reduced levels of physical risk to their soldiers. For better or worse, the use of military robots will minimise the psychological aspects of combat stress. They may also prove to reduce the environmental impact of military forces and, more importantly, are cheaper and more effective than human warfighters. In short, there are great military, strategic and political rewards of moral relevance that are associated with the use of these machines. Chapter 4 sought to provide a broad but complete contractually grounded version of just war theory and explored the restraint/s that should come into play in the search for the benefits that accompany unmanned systems. In response to critics levelling the claim that just war theory holds little contemporary relevance, it has been suggested that interpreting and applying existing just war principles to unmanned warfare is bound to be fraught with complexity, but represents our best chance at tempering states' responses to perceived threats.

The remaining chapters, which constitute the core of this work, detailed these complexities. Chapter 5 addressed the potential technological and operational problems associated with the employment of unmanned systems, with particular fears about complacency with human life and technical limitations, especially in our current risk climate in which American and allied forces are not to be put in harm's way. Chapter 6, on the other hand, addressed socio-political and psychological factors. In many cases, would-be targets, other potential victims, citizens and external observers are unaware or hold manipulated views of drone strikes, calling into question adherence to the principle of right authority and others seeking to govern justice in war. Some believe that there are further technological solutions to these problems but these, coupled with the fact that world governments anticipate a need to increase the scale and intensity of their drone usage, will further the problems of radical asymmetry described in Chapter 7. The concern here is that some states may eventually come to wield such strength in military robotics, with risks to the state that are so low that it would present an impediment to just war. Chapter 8 ventured into the unexplored territory that is war's end and explained the role that unmanned systems can play in facilitating a lasting peace if used to support transparent communication and directly aid support operations, rather than angering and scaring the local populace.

The message put forward here, then, is that each step we take toward making the West safer and more prosperous through the use of innovative military robotics is potentially dangerous for all the relevant stakeholders. The other line of argument running throughout this work is that we ought to be sceptical of any narrative that is overly optimistic or cynical. Tales of the overly optimistic variety are often the first to attract critique, but stories of the overly cynical variety can also steer the debate in the wrong direction. For instance, some anti-robot consortia are concerned that there is an absence of clear scientific evidence that highly autonomous unmanned systems have, or are likely to have in the foreseeable future, the functionality required for accurate target identification, situational awareness or decisions regarding the proportional use of force. Thus, they think that autonomous systems will undermine human responsibility, and will be unable to meet the strict requirements for the use of force. They conclude that if this is the case, then the use of autonomous unmanned systems should be banned. This is especially true, they argue, under modern conditions in which battlefields are not clearly delimited and discrimination between civilians, insurgents and combatants is increasingly difficult.

As described in Chapter 9, there is indeed a range of complex problems associated with attributing responsibility. However, neither this problem relating to responsibility, nor any of the other problems summarised in this book, offers a sufficient justification for a complete prohibition on the development and deployment of autonomous unmanned systems when taken in isolation. The task before us is to determine whether we can adequately manage and transform the relevant challenges into a useful operational solution, rather than prematurely arguing for an outright ban. The first step we ought to take towards this goal is to acknowledge that unmanned warfare is a human activity and social institution, despite the fact that it aims to reduce or eliminate soldiers from the battlefield. Recall that war was defined as organised violence maintained through contractual relations. Human beings are the principal drivers of unmanned war and they define their options and make choices for reasons that are psychological, social and cultural in nature. Going forward, we must urgently seek to understand these influences on an empirical level and then further consider their politico-strategic, legal and moral import. While it is safe to say that the moral landscape is murky, there would seem to be some legitimate roles for even the most autonomous systems. For instance, there is no obvious moral reason to prohibit their use for some very specific purposes, whether it be patrolling borders in remote areas or taking part in battles within the confines of a 'kill box' in which no civilians stand to be harmed. Realistic regulation, however, is very much necessary.

References

Abdul-Ahad, G. 'Yemenis Choose Jihad over Iranian Support'. In *The Guardian* (2012). Published electronically 10 May. http://www.guardian.co.uk/world/2012/may/10/yemen-jihad-iran-saudi-interference.

Ackerman, G. A. 'It Is Hard to Predict the Future: The Evolving Nature of Threats and Vulnerabilities'. *Revue Scientifique et Technique International Office of Epizootics* 25, no. 1 (2006): 353–60.

Adams, T. 'Future Warfare and the Decline of Human Decisionmaking'. *Parameters: The US Army War College Quarterly Senior Professional Journal* 31, no. 4 (2001): 57–71.

Alford, H. *The Works of John Donne, with a Memoir of His Life*. London: John W. Parker, 1839.

Allen, N. 'Just War in the Mahabharata'. In *The Ethics of War: Shared Problems in Different Traditions*, edited by R. Sorabji and D. Rodin, 138–49. Aldershot: Ashgate, 2006.

Allison, R. 'Russia, Regional Conflict, and the Use of Military Power'. In *The Russian Military: Power and Policy*, edited by S. E. Miller and D. Trenin. 121–56. Massachusetts: MIT Press, 2004.

Anderson, S., K. Madhava Sarma and K. Taddonio. *Technology Transfer for the Ozone Layer: Lessons for Climate Change*. London: Earthscan, 2007.

Anscombe, E. 'War and Murder'. In *War, Morality, and the Military Profession*, edited by M. M. Wakin. 284–316. Boulder, CO: Westview Press, 1986.

Arkin, R. 'Ethical Robots in Warfare'. *IEEE Technology and Society* 28, no. 1 (2009): 30–33.

Arkin, R. 'Governing Lethal Behaviour'. In *Ethical and Legal Aspects of Unmanned Systems: Interviews*, edited by Gerhard Dabringer. 149–56. Vienna: Institute for Religion and Peace, 2010.

Arkin, R. *Governing Lethal Behavior in Autonomous Robots*. Boca Raton: CRC Press, 2009.

Arkin, R. 'On the Ethical Quandries of a Practicing Roboticist: A First-Hand Look'. In *Current Issues in Computing and Philosophy*, edited by A. Briggle, K. Waelbers and P. Brey. 45–49. Amsterdam: IOS Press, 2008.

Arkin, R. 'The Case for Ethical Autonomy in Unmanned Systems'. *Journal of Military Ethics* 9, no. 4 (2010): 332–41.

Asaro, P. M. 'How Just Could a Robot War Be?' In *Current Issues in Computing and Philosophy*, edited by P. Brey, A. Briggle and K. Waelbers. 50–64. Amsterdam: IOS Press, 2008.

Asaro, P. M. 'Military Robots and Just War Theory'. In *Ethical and Legal Aspects of Unmanned Systems: Interviews*, edited by G. Dabringer. 103–19. Vienna: Institute for Religion and Peace, 2010.

Asaro, P. M. 'Modeling the Moral User'. *IEEE Technology and Society* 28, no. 1 (2009): 20–24.

Associated Press. 'Text of Bush Speech'. CBS News, http://www.cbsnews.com/stories/2003/05/01/iraq/main551946.shtml.

Associated Press. 'US Military Damaged Ancient Babylon Site'. Sydney Morning Herald, http://www.smh.com.au/world/us-military-damaged-ancient-baby lon-site-20090710-deuh.html.

Augustine. *City of God*. London: Penguin Books, 1984.

Axe, D. *War Bots: How U.S. Military Robots Are Transforming War in Iraq, Afghanistan, and the Future*. Ann Arbor: Nimble Books, 2008.

Bacevich, A. J. *The New American Militarism: How Americans Are Seduced by War*. Oxford: Oxford University Press, 2005.

BAE Systems. 'BAE Systems Unveils Latest Innovative Unmanned Underwater System'. http://www.baesystems.com/Newsroom/NewsReleases/autoGen_107227114948.html.

Baker, D. P. *Just Warriors Inc: The Ethics of Privatised Force*. New York: Continuum, 2011.

Baker, D. P. 'To Whom Does a Private Military Commander Owe Allegiance?' In *New Wars and New Soldiers: Military Ethics in the Contemporary World*, edited by P. Tripodi and J. Wolfendale. 181–200. Farnham: Ashgate, 2011.

Balyko, Y. 'NATO's Unmanned Aerial Vehicles in Local Conflicts'. *Military Parade* 1 (2008): 36–37.

Barker, J. *Agincourt: The King, the Campaign, the Battle*. London: Little & Brown, 2005.

Barnes, M. and A. W. Evans. 'Soldier-Robot Teams in Future Battlefields: An Overview'. In *Human-Robot Interactions in Future Military Operations*, edited by M. Barnes and F. Jentsch. 9–30. Barnes: Ashgate, 2011.

Barnes, M., H. R. Everett and P. Rudakevych. 'Throwbot: Design Considerations for a Man-Portable Throwable Robot'. Paper presented at the SPIE, March 2005.

Barrett, E. 'Executive Summary and Command Brief'. *Journal of Military Ethics* 9, no. 4 (2010): 424–31.

Bartlett, R. *United States Navy*. London: Heinemann Library, 2004.

Bash, H. H. *Stuff 'N' Nonsense: A Hodge-Podge of Whimsical, Sober & Politically Incorrect Musings*. Toronto: Inkwater Press, 2011.

Baumgold, D. 'Subjects and Soldiers: Hobbes on Military Service'. *History of Political Thought* 4, no. 1 (1983): 43–63.

Beauchamp, Z. and J. Savulescu. 'Robot Guardians: Teleoperated Combat Vehicles in Humanitarian Military Intervention'. In *Killing by Remote Control: The Ethics of an Unmanned Military*, edited by B. J. Strawser. 106–25. New York: Oxford, 2013.

Beck, U. *World Risk Society*. Cambridge: Polity, 1999.

Belin, D. and G. Chapman. *Computers in Battle: Will They Work?* New York: Harcourt Brace Jovanovich Publishers, 1987.

Bellamy, A. J. 'The Responsibilities of Victory: Jus Post Bellum and the Just War'. *Review of International Studies* 34, no. 4 (2008): 601–25.

Benjamin, M. *Drone Warfare: Killing by Remote Control*. New York: OR Books, 2012.

Bergen, P. and K. Teidmann. 'The Year of the Drone: An Analysis of U.S. Drone Strikes in Pakistan 2004–2010'. *Foreign Policy* (2010). Published electronically 26 April. http://www.foreignpolicy.com/articles/2010/04/26/the_year_of_the_drone.

Bergen, P. and K. Teidmann. 'Washington's Phantom War: The Effects of the U.S. Drone Program in Pakistan'. *Foreign Affairs* 90, no. 4 (2011): 12–18.

Betz, D. J. and S. Lee. 'Information in the Western Way of Warfare-Too Much of a Good Thing?' *Pacific Focus* 21, no. 2 (2006): 197–231.

Biddle, S. *Afghanistan and the Future of Warfare: Implications for Army and Defense Policy*. Carlisle: Strategic Studies Institute, 2002.

Black, J. 'The Revolution in Military Affairs: The Historian's Perspective'. In *Journal of Military and Strategic Studies* 9, no. 2 (2006/7): 1–42. www.jmss.org/jmss/index.php/jmss/article/download/111/122.

Bonds, R. *Illustrated Directory of Modern American Weapons*. Minneapolis: Zenith Imprint, 2002.

Booher, L. and J. Appezzato. 'What Could Happen? No Room for Complacency in OIF'. *Combat Edge* 13, no. 10 (2005): 4–7.

Bookstaber, R. 'Will the Unemployed Really Find Jobs Making Robots?' http://www.businessinsider.com/will-the-unemployed-really-find-jobs-making-robots-2012-8.

Boot, M. *War Made New: Technology, Warfare, and the Course of History, 1500 to Today*. New York: Gotham Books, 2006.

Bosanquet, C. 'The Ethics of a Just Post Bellum'. In *International Society of Military Ethics*, 2007.

Bowyer, D. 'Just War Doctrine: Relevance and Challenges in the 21st Century'. Doctor of Philosophy, Cranfield University, 2008.

Boyne, W. J. *How the Helicopter Changed Modern Warfare*. Gretna: Pelican Publishing, 2011.

Brandt, R. 'Utilitarianism and the Rules of War'. In *War, Morality, and the Military Profession*, edited by M. Wakin. 393–413. Westview Press: Boulder, 1981.

Braun, M. and D. Brunstetter. 'Rethinking the Criterion for Assessing CIA-Targeted Killings: Drones, Proportionality and Jus Ad Vim'. *Journal of Military Ethics* 12, no. 4 (2014): 304–24.

Brennan, J. O. 'The Efficacy and Ethics of U.S. Counterterrorism Strategy'. Woodrow Wilson International Center for Scholars, 2012.

British Broadcasting Corporation. 'Belgrade Hit by Nato Blitz'. http://news.bbc.co.uk/2/hi/europe/332328.stm.

Bronner, E. 'Israel Plans New Rules on Exchange of Prisoners'. *The New York Times* (2012). http://www.nytimes.com/2012/01/06/world/middleeast/after-shalit-israel-changing-prisoner-exchange-rules.html.

Bruneau, T. C. 'Civil-Military Relations in Latin America: The Hedgehog and the Fox Revisited'. *Revista Fuerzas Armadas y Sociedad* 19, no. 1 (2005): 111–31.

Bruneau, T. C. *Patriots for Profit: Contractors and the Military in the U.S. National Security*. Stanford: Stanford University Press, 2011.

Bruneau, T. C. and F. C. Matei. 'Towards a New Conceptualization of Democratization and Civil-Military Relations'. *Democratization* 15, no. 5 (2008): 909–29.

Brunstetter, D. and M. Braun. 'The Implications of Drones on the Just War Tradition'. *Ethics & International Affairs* 25, no. 3 (2011): 337–58.

Bryant, J. K. *The Battle of Fredericksburg: We Cannot Escape History*. Charleston: The History Press, 2010.

Bryce, M. 'The Unique Nature of Military Service'. In *Defence Force Welfare Association Forum*. Canberra: Defence Force Welfare Association, 2009.

Buley, B. *The New American Way of War: Military Culture and the Political Utility of Force*. New York: Routledge, 2008.

Burgan, M. *Nikola Tesla: Physicist, Inventor, Electrical Engineer*. Mankakto: Compass Point Books, 2009.

Cady, D. L. *From Warism to Pacifism: A Moral Continuum*. Philadelphia: Temple University Press, 2010.

Calhoun, L. 'Political Philosophers on War: Arguments inside the "Just War" Box'. *The Independent Review* 15, no. 3 (2011): 447–61.

Cameron, L. 'Private Military Companies: Their Status under International Law and Its Impact on Their Regulation'. *International Review of the Red Cross* 88, no. 863 (2006): 573–98.

Campbell, R. H. and G. Eaton. 'It's All About You: Implementing Human-Centered Design'. In *Human Performance Enhacement in High-Risk Environments: Insights, Developments, and Future Directions from Military Research*, edited by P. O'Connor and J. Cohn. 21–223. Santa Barbara: ABC-CLIO, 2011.

Canning, J. and G. Dabringer. 'Ethical Challenges of Unmanned Systems'. In *Ethical and Legal Aspects of Unmanned Systems: Interviews*, edited by G. Dabringer. 7–19. Vienna: Institute for Religion and Peace, 2010.

Capek, K. *R.U.R (Rossum's Universal Robots)*, translated by D. Wyllie. Fairford: Echo Library, 2010.

Cavelty, M. D. 'Cyberwar'. In *The Ashgate Research Companion to Modern Warfare*, edited by George Kassimeris and George Buckley. Farnham: Ashgate, 2010.

Center for Civilians in Conflict. 'The Civilian Impact of Drones: Unexamined Costs, Unanswered Questions'. New York: Columbia Law School Human Rights Clinic, 2012.

Chappelle, W., K. McDonald, and K. McMillan. 'Important and Critical Psychological Attributes of USAF MQ-1 Predator and MQ-9 Reaper Pilots According to Subject Matter Experts', edited by Department of the Airforce. Ohio: School of Aerospace Medicine, 2011.

Childress, J. F. 'Just-War Theories: The Bases, Interrelations, Priorities, and Functions of Their Criteria'. *Theological Studies* 39, no. 1 (1978): 427–45.

Chiu, Y. 'Combat Contractualism: A Secular Theory of Just War'. University of California, 2007.

Chow, J. C. 'The Case for Humanitarian Drones'. *Open Canada* (2012). http://opencanada.org/features/the-think-tank/essays/the-case-for-humanitarian-drones/.

Christopher, P. *The Ethics of War and Peace.* Upper Saddle River, NJ: Prentice Hall, 1994.

Churchill, W. *The River War: An Historical Account of the Reconquest of the Soudan.* London: Longmans, Green and Co., 1899.

Clark, R. *The Role of the Bomber.* New York: Thomas Y. Cromwell Company, 1977.

Clement, T. 'The Morality and Economics of Safety in Defence Procurement'. In *Safety-Critical Systems: Problems, Process and Practice*, edited by C. Dale and T. Anderson. 39–54. Dordrecht: Springer, 2009.

Coady, C. A. J. *Morality and Political Violence.* Cambridge: Cambridge University Press, 2008.

Coady, C. A. J. 'The Jus Post Bellum'. In *New Wars and New Soldiers: Military Ethics in the Contemporary World*, edited by P. Tripodi and J. Wolfendale. 49–58. Farnham: Ashgate, 2011.

Coeckelbergh, M. 'From Killer Machines to Doctrines and Swarms, or Why Ethics of Military Robotics Is Not (Necessarily) About Robots'. *Philosophy & Technology* 24, no. 3 (2011): 269–78.

Coker, C. *Ethics and War in the 21st Century.* New York: Routledge, 2008.

Coker, C. *The Warrior Ethos: Military Culture and the War on Terror.* New York: Routledge, 2007.

Coker, C. *Waging War without Warriors? The Changing Culture of Military Conflict.* Boulder, CO: Lynne Rienner Publishers, 2002.

Coleman, S. *Military Ethics: An Introduction with Case Studies.* Oxford: Oxford University Press, 2012.

Congressional Unmanned Systems Caucus. http://unmannedsystemscaucus.mckeon.house.gov/about/purpose-mission-goals.shtml.

Cook, M. *The Moral Warrior: Ethics and Service in the U.S. Military.* Albany: State University of New York Press, 2004.

Cooke, N. J. and R. A. Chadwick. 'Lessons Learned from Human-Robotic Interactions on the Ground and in the Air'. In *Human-Robot Interactions in Future Military Operations*, edited by M. Barnes and F. Jentsch. 355–74. Farnham: Ashgate, 2011.

Cooper, S. 'High-Tech War and the Securing of Consent'. *Southern Review: Communication, Politics & Culture* 37, no. 3 (2005): 73–86.

Coppieters, B. and B. Kashnikov. 'Right Intentions'. In *Moral Constraints on War: Principles and Cases*, edited by B. Coppieters and N. Fotion. 73–101. Lanham: Lexington Books, 2008.

Coppieters, B., R. Apressyan, and C. Ceulemans. 'Last Resort'. In *Moral Constraints on War: Principles and Cases*, edited by B. Coppieters and N. Fotion. 139–54. Lanham: Lexington Books, 2008.

Coppieters, B., C. Ceulemans, and A. E. Hartle. 'Just Cause'. In *Moral Constraints on War: Principles and Cases*, edited by B. Coppieters and N. Fotion. 27–54. Lanham: Lexington Books, 2008.

Cordesman, A. *The Iraq War: Strategy, Tactics, and Military Lessons.* Washington, DC: CSIS Press, 2003.

Cordesman, A. 'The Lessons and Non-Lessons of the Air and Missile War in Kosovo'. Washington, DC: Center for Strategic and International Studies, 1999.

Corum, J. S. 'Development of Modern Counterinsurgency Theory and Doctrine'. In *The Ashgate Research Companion to Modern Warfare*, edited by G. Kassimeris and G. Buckley. Farnham: Ashgate, 2010.

Cosenzo, K. 'Automation Strategies for Facilitating Human Interaction with Military Unmanned Vehicles'. In *Human-Robot Interactions in Future Military Operations*, edited by M. Barnes and F. Jentsch. Farnham: Ashgate, 2011.

Crabtree, J. *On Air Defense.* Westport: Greenwood Publishing, 1994.

Craik, K. J. W. 'Theory of the Human Operator in Control Systems I: The Operator as an Engineering System'. *British Journal of Psychology* 38, no. 2 (1948): 142–48.

Craik, K. J. W. 'Theory of the Human Operator in Control Systems II: Man as an Element in a Control System'. *British Journal of Psychology* 38, no. 3 (1948): 142–48.

Croome, A. *Midnight Empire.* Sydney: Allen & Unwin, 2012.

Cummings, M. L. 'Automation and Accountability in Decision Support System Interface Design'. *Journal of Technology Studies* 32, no. 1 (2006): 23–31.

Cummings, M. L. 'Automation Bias in Intelligent Time Critical Decision Support Systems'. In *AIAA 1st Intelligent Systems Technical Conference*. Chicago, 2004.

Cummings, M. L. 'Integrating Ethics in Design through the Value-Sensitive Design Approach'. *Science and Engineering Ethics* 12, no. 4 (2006): 701–15.

Cummins, J. *Why Some Wars Never End: The Stories of the Longest Conflicts in History*. Massachusetts: Fair Winds Press, 2010.

Daniel, L. 'Military Leads Mental Health Care Transformation'. *American Forces Press Service*, 24 February 2010.

Darling, K. 'Extending Legal Rights to Social Robots'. In *We Robot*. Coral Gables: University of Miami School of Law, 2012.

Davis, L. *Natural Disasters*. New York: Facts on File, Inc., 2008.

DeLanda, M. *War in the Age of Intelligent Machines*. New York: Swerve Editions, 1991.

Dennett, D. D. 'Mechanism and Responsibility'. In *Essays on Freedom of Action*, edited by T. Honderich. Boston: Routledge and Keegan Paul, 1973.

Dennett, D. D. 'When HAL Kills, Who's to Blame? Computer Ethics'. In *HAL's Legacy: 2001's Computer as a Dream and Reality*, edited by D. G. Stork. Cambridge: MIT Press, 1997.

Didziulis, J. R. 'Winning the Battle for Hearts and Minds: Operationalizing Cultural Awareness During Stability Operations'. Montgomery: Air Command and Staff College, 2008.

Dipert, R. R. 'The Ethics of Cyberwarfare'. *Journal of Military Ethics* 9, no. 4 (2010): 384–410.

Dobos, N. 'Are Our Soldiers Assets or Workers?' *Sydney Morning Herald*, 4 June 2012.

Dodd, M. 'Combat Roles Offered to Women'. *The Australian*, 12 April 2011.

DoDvClips. 'UAV Strike on Insurgents'. http://www.youtube.com/watch?v=QZ-dNu5uOQc&feature=related.

Dubber, M. D. *Victims in the War on Crime: The Use and Abuse of Victims' Rights*. New York: New York University Press, 2002.

Dunlap, C. 'Organizational Change and the New Technologies of War'. In *Joint Services Conference on Professional Ethics*. Washington, DC, 1998.

Dunlap, C. *Technology and the 21st Century Battlefield: Recomplicating Moral Life for the Statesman and the Soldier*. Carlisle: Institute for Strategic Studies, 1999.

Dyer, G. *War: The New Edition*. New York: Random House, 2010.

Egeland, J. *A Billion Lives: An Eyewitness Report from the Frontlines of Humanity*. New York: Simon & Schuster, 2008.

Ehrhard, T. P. *Air Force UAVs: The Secret History*. Washington, DC: Mitchell Institute Press, 2010.

Eisenhower, D. D. *Dwight D. Eisenhower: 1960–61: Containing the Public Messages, Speeches, and Statements of the President, January 1, 1960, to January 20, 1961*. Washington, DC: United States Government Printing Office, 1961.

Ely, J. A. 'Drones: A Challenge to the Professional Military Ethic'. https://globalecco.org/drones-a-challenge-to-the-professional-military-ethic.

Enemark, C. *Armed Drones and the Ethics of War: Military Virtue in a Post-Heroic Age*. New York: Routledge, 2013.

Enemark, C. 'Drones over Pakistan: Secrecy, Ethics, and Counterinsurgency'. *Asian Security* 7, no. 3 (2011): 218–37.

Evans, M. 'In Defence of Just War Theory'. In *Just War Theory: A Reappraisal*, edited by Mark Evans. 203–22. Edinburgh: Edinburgh University Press, 2005.

Fabre, C. 'Cosmopolitanism, Just War Theory and Legitimate Authority'. *Journal of International Affairs* 84, no. 5 (2008): 963–76.

Fahlquist, J. N. 'Moral Responsibility for Environmental Problems – Individual or Institutional?' *Journal of Agricultural and Environmental Ethics* 22, no. 2 (2008): 109–24.

Feaver, P. D. *Armed Servants: Agency, Oversight, and Civil-Military Relations*. Cambridge: Harvard University Press, 2003.

Feaver, P. D. 'Civil-Military Relations'. *Annual Review of Political Science* 2, no. 1 (1999): 211–41.

Fiala, A. *Practical Pacifism*. New York: Algora Publishing, 2004.

Fiala, A. *The Just War Myth*. Lanham: Lexington Books, 2008.

Fieser, J. and B. Dowden. 'Just War Theory'. *The Internet Encyclopedia of Philosophy* (2007). http://www.iep.utm.edu/j/justwar.htm.

Finlan, A. *The Gulf War of 1991*. New York: Rosen Publishing Group, 2008.

Finn, A. and S. Scheding. *Developments and Challenges for Autonomous Unmanned Vehicles: A Compendium*. Berlin: Springer Verlag, 2010.

Finnegan, J. *Military Intelligence: A Picture History*. Arlington: United States Army Intelligence and Security Command, 1985.

Fischer, J. M. and M. Ravizza, eds. *Perspectives on Moral Responsibility*. Ithaca: Cornell University Press, 1993.

Fischer, J. M. and M. Ravizza, eds. *Responsibility and Control: A Theory of Moral Responsibility*. Cambridge: Cambridge University Press, 1998.

Fisher, D. *Morality and War: Can War Be Just in the Twenty-First Century?* Oxford: Oxford University Press, 2011.

Fitzsimonds, J. R. and T. G. Mahnken. 'Military Officer Attitudes toward UAV Adoption: Exploring Institutional Impediments to Innovation'. *Joint Forces Quarterly* 46, no. 3 (2007): 96–103.

Floridi, L. 'Children of the Fourth Revolution'. *Journal of Philosophy and Technology* 24, no. 3 (2011): 227–32.

Floridi, L. and J. Sancers. 'The Foundationalist Debate in Computer Ethics'. In *Readings in Cyberethics*, edited by R. Spinello and H. Tavani. 81–95. Massachusetts: Jones and Bartlett, 2004.

Fotion, N. 'A Utilitarian Defense of Just War Theory'. *Synthesis Philosophica* 12, no. 1 (1997): 209–25.

Fotion, N. 'Proportionality'. In *Moral Constraints on War: Principles and Cases*, edited by B. Coppieters and N. Fotion. 125–37. Lanham: Lexington Books, 2008.

Fotion, N. and B. Coppieters. 'The Likelihood of Success'. In *Moral Constraints on War: Principles and Cases*, edited by B. Coppieters and N. Fotion. 101–24. Lanham: Lexington Books, 2008.

Fotion, N., B. Coppieters, and R. Apressyan. 'Introduction'. In *Moral Constraints on War: Principles and Cases*, edited by B. Coppieters and N. Fotion. 1–24. Lanham: Lexington Books, 2008.

Francis, P. L. 'Issues to Be Considered for Army's Modernization of Combat Systems', edited by United States Government Accountability Office, 1–18. Washington, DC: United States Government Accountability Office, 2009.

Friedman, B. 'Moral Responsibility and Computer Technology'. In *Annual Meeting of the American Educational Research Association*. Boston, MA, 1990.

Friedman, R. S. *Advanced Technology Warfare: A Detailed Study of the Latest Weapons and Techniques for Warfare Today and into the 21st Century*. New York: Harmony Books, 1985.

Gabriel, R. A. and K. S. Metz. *From Sumer to Rome: The Military Capabilities of Ancient Times*. New York: Greenwood Press, 1991.

Gage, D. W. 'UGV History 101: A Brief History of Unmanned Ground Vehicle (UGV) Development Efforts'. *Unmanned Systems Magazine* 13, no. 3 (1995): 9–16.

Galliott, J. C. 'Closing with Completeness: The Asymmetric Drone Warfare Debate'. *Journal of Military Ethics* 11, no. 4 (2012/13): 353–56.

Galliott, J. C., ed. *Commercial Space Exploration: Ethics, Policy and Governance*. Farnham: Ashgate (forthcoming).

Galliott, J. C. 'Take out the Pilot from Australia's Joint Strike Fighter'. *The Conversation* (2014). https://theconversation.com/take-out-the-pilot-from-australias-f-35-joint-strike-fighter-28107.

Galliott, J. C. 'Uninhabited Systems and the Asymmetry Objection: A Response to Strawser'. *Journal of Military Ethics* 11, no. 1 (2012): 58–66.

Gardner, L. 'America and the Vietnam War: Re-Examining the Culture and History of a Generation'. In *Iraq as the "Good War" as Opposed to Vietnam, the Bad War*, edited by A. A. Wiest, M. K. Barbier and G. Robins. 291–311. New York: Taylor & Francis, 2009.

Garren, D. J. 'Soldiers, Slaves and the Liberal State'. *Philosophy & Public Policy Quarterly* 27, no. 1/2 (2007): 8–11.

Garvey, J. *The Ethics of Climate Change: Right and Wrong in a Warning World*. New York: Bloomsbury, 2008.

Gere, F. 'The Future of Asymmetric Warfare'. In *The Ashgate Research Companion to War: Origins and Prevention*, edited by Hall Gardner and Oleg Kobtzefff. 505–28. Farnham: Ashgate, 2012.

Gertz, N. 'Technology and Suffering in War'. In *Technology and Security*. University of North Texas: Society for Philosophy and Technology, 2011.

Gideon, F. C. 'Air Force Instruction 90–901'. Washington, DC: United States Department of Defense, 2000.

Glosson, B. C. 'Impact of Precision Weapons on Air Combat Operations'. *Airpower Journal* 7, no. 2 (1993): 4–10.

Goncalves, R., S. Ferreira, J. Pinto, et al. 'Authority Sharing in Mixed Initiative Control of Mulitple Uninhabited Aerial Vehicles'. Paper presented at the Engineering Psychology and Cognitive Ergonomics, Orlando, Florida, 2011.

Goodrick, M. A. 'On Maximising Fan-Out: Towards Controlling Multiple Unmanned Vehicles'. In *Human-Robot Interactions in Future Military Operations*, edited by M. Barnes and F. Jentsch. Farnham: Ashgate, 2011.

Gotterbarn, D. 'Informatics and Professional Responsibility'. *Science and Engineering Ethics* 7, no. 2 (2001): 221–30.

Graham, S. 'The New Military Urbanism'. In *The New Blackwell Companion to the City*, edited by G. Bridge and S. Watson. 121–33. Malden: Wiley-Blackwell, 2011.

Gray, C. S. 'AI at War: The Aegis System in Combat'. In *Directions and Implications of Advanced Computing*, edited by D. Shculer. 62–79. New York: Ablex, 1997.

Gray, C. S. *Another Bloody Century: Future Warfare*. London: Weidenfeld & Nicolson, 2005.

Gray, C. S. 'The 21st Century Security Environment and the Future of War'. *Parameters: The US Army War College Quarterly Senior Professional Journal* 38, no. 4 (2008): 14–26.

Gray, G. J. *The Warriors: Reflections on Men in Battle*. New York: Harper & Row, 1959.

Greek Elegiac Poetry: From the Seventh to the Fifth Centuries B.C., translated by D. E. Gerber. Cambridge: Harvard University Press, 1999.

Grossman, D. *On Killing: The Psychological Cost of Learning to Kill in War and Society*. Boston: Little, Brown and Company, 1995.

Grotius, H. 'On the Law of War and Peace: Three Books'. In *The Classics of International Law*, edited by J. B. Scott Washington, DC: Carnegie Institution of Washington, 1925.

Grotius, H. *The Rights of War and Peace*, translated by A. C. Campbell. Vol. 2, 24. Washington, DC: M. Walter Dunn, 1901.

Hall, J. *Low Reynolds Number Aerodynamics for Micro Aerial Vehicles*. Boulder, CO: University of Colorado, 2007.

Halverson, N. 'Marine Corps Orders Bulletproof Underwear'. Discovery Channel. http://news.discovery.com/tech/marine-corps-orders-bulletproof-underwear-110418.html.

Hanson, D. 'Predator Strike Video'. http://www.youtube.com/watch?v=tIsdwbZSMMw.

Hare, R. M. 'Rules of War and Moral Reasoning'. In *Essays on Political Morality*, edited by R. M. Hare. 45–61. Oxford: Clarendon Press, 1989.

Hartle, A. E. 'Discrimination'. In *Moral Constraints on War: Principles and Cases*, edited by Bruno Coppieters and Nick Fotion. 171–90. Lanham: Lexington Books, 2008.

Hauck, L. 'Have We Become … Complacent?' *Combat Edge* 11, no. 11 (2003): 8–11.

Hauert, S., S. Leven, J. -C. Zuffrey, and D. Floreano. 'The Swarming Micro Air Vehicle Network (SMAVNET) Project'. http://lis2.epfl.ch/CompletedResearchProjects/SwarmingMAVs/.

Hay-Edie, D. *The Military's Impact on the Environment: A Neglected Aspect of the Sustainable Development Debate.* Geneva: International Peace Bureau, 2002.

Headrick, D. R. *Power over Peoples: Technology, Environments, and Western Imperialism.* New Haven: Princeton University Press, 2010.

Held, V. *How Terrorism Is Wrong: Morality and Political Violence.* Oxford: Oxford University Press, 2008.

Hennigan, W. J. 'It's a Bird! It's a Spy! It's Both'. *Los Angeles Times.* http://articles.latimes.com/2011/feb/17/business/la-fi-hummingbird-drone-20110217.

Herspring, D. R. *The Pentagon and the Presidency: Civil-Military Relations from FDR to George W. Bush.* Lawrence: University of Kansas Press, 2005.

Hew, P. 'C2 Design for Ethical Agency over Killing in War'. In *The Evolution of C2: Where Have We Been? Where Are We Going?* Santa Monica: International Command and Control Research and Technology Symposia, 2010.

Hew, P. 'The Blind Spot in Robot-Enabled Warfare Deeper Implications of the IED Challenge'. *Australian Army Journal* 7, no. 2 (2010): 45–56.

Hills, A. *Future War in Cities: Rethinking a Liberal Dilemma.* Portland: Frank Cass, 2004.

Hobbes, T. *Leviathan.* Sioux Falls: Nuvision Publishers, 2004.

Homer. *The Iliad and Odyssey of Homer*, translated by A. Pope, edited by H. F. Cary London: George Routledge and Sons, 1890.

Homer. *The Iliad of Homer*, translated by R. Lattimore. Chicago: University of Chicago Press, 1951.

Homer. *The Odyssey*, translated by W. Shewring. Oxford: Oxford University Press, 1980.

Horner, D. 'Stress on Higher Commanders in Future Warfare'. In *The Human Face of Warfare: Killing, Fear and Chaos in Battle*, edited by M. Evans and A. Ryan. 134–58. St Leonards: Allen & Unwin, 2000.

Hoskins, A. *Televising War: From Vietnam to Iraq.* London: Continuum International Publishing Group, 2004.

House of Commons Foreign Affairs Committee. 'Global Security: Afghanistan and Pakistan, Eighth Report of Session 2008–09, Report, Together with

Formal Minutes, Oral and Written Evidence'. London: Parliament: House of Commons: Foreign Affairs Committee, 2009.

Hudson, L., C. Owens, and M. Flannes. 'Drone Warfare: Blowback from the American Way of War'. *Middle East Policy* 18, no. 3 (2011): 122–32.

Hunt, D. V. *Smart Robots: A Handbook of Intelligent Robot Systems*. Dordrecht: Springer, 1985.

Huntington, S. P. *The Soldier and the State: The Theory and Politics of Civil-Military Relations*. Cambridge: Belknap Press, 1981.

Hurka, T. 'Proportionality in the Morality of War'. *Philosophy & Public Affairs* 33, no. 1 (2004): 34–66.

Ignatieff, M. *The Warrior's Honour: Ethnic War and the Modern Conscience*. London: Vintage, 1998.

Ignatieff, M. *Virtual War: Kosovo and Beyond*. New York: Metropolitan Books, 2000.

Institute of Medicine. *Gulf War and Health: Depleted Uranium, Sarin, Pyridostigmine*. Washington, DC: National Academy Press, 2000.

International Committee for Robot Arms Control. 'Scientists's Call'. http://icrac.net/call/.

Jalbert, J. 'Solar-Powered Autonomous Underwater Vehicle Development'. Paper presented at the 13th International Symposium on Unmanned Untethered Submersible Technology, Durham, August 2003.

Johnson, D. D. P. *Overconfidence and War: The Havoc and Glory of Positive Illusions*. Cambridge: Harvard University Press, 2004.

Johnson, J. T. *Can Modern War Be Just?* New Haven: Yale University Press, 1984.

Johnson, J. T. *Ethics and the Use of Force: Just War in Historical Perspective*. Farnham: Ashgate, 2011.

Johnson, J. T. *Just War Tradition and the Restraint of War*. Princeton: Princeton University Press, 1981.

Johnson, J. T. 'Moral Responsibility after Conflict'. In *Ethics Beyond War's End*, edited by E. D. Patterson, 17–34. Washington, DC: Georgetown University Press, 2012.

Johnson, R. J. 'The Wizard of Oz Goes to War: Unmanned Systems in Counterinsurgency'. In *Killing by Remote Control: The Ethics of an Unmanned Military*, edited by B. J. Strawser. 154–78. New York: Oxford, 2013.

Jonas, H. *The Imperative of Responsibility. In Search of an Ethics for the Technological Age*. Chicago: University of Chicago Press, 1984.

Kahn, P. 'The Paradox of Riskless Warfare'. *Philosophy and Public Policy Quarterly* 22, no. 3 (2002): 2–8.

Kaltenhaler, K., W. Miller, and C. Fair. 'The Drone War: Pakistani Public Attitudes toward American Drone Strikes in Pakistan'. In *Annual Meeting of the Midwest Political Science Association*. Chicago, 2012.

Kant, I. 'Metaphysics of Morals'. Translated by H. B. Nisbet. In *Kant: Political Writings*, edited by H. S. Reiss. Cambridge: Cambridge University Press, 1991.

Kant, I. *The Philosophy of Law: An Exposition of the Fundamental Principles of Jurisprudence as the Science of Right.* Translated by W. Hastie. Edinburgh: T. & T. Clark, 1887.

Kasachkoff, T. and J. Kleinig. 'Drones, Distance and Death'. In *The Rule of Law in an Era of Change: Security, Social Justice and Inclusive Government.* Athens, Greece, 2014.

Kasher, A. and A. Plaw. 'Distinguishing Drones: An Exchange'. In *Killing by Remote Control: The Ethics of an Unmanned Military*, edited by B. J. Strawser. 47–68. New York: Oxford, 2013.

Kaurin, P. 'Courage Behind a Screen'. 2013.

Keegan, J. and R. Holmes. *Soldiers: A History of Men in Battle.* New York: Viking, 1986.

Keller, B. 'Smart Drones'. *The New York Times* (2013). http://www.nytimes.com/2013/03/17/opinion/sunday/keller-smart-drones.html?pagewanted=all.

Kels, C. G. 'Closing Argument: At the Outer Bounds of Asymmetry'. *Journal of Military Ethics* 11, no. 3 (2012): 223–44.

Kemp, G. 'Arms Acquisition and Violence: Are Weapons or People the Cause of Conflict?' In *Leashing the Dogs of War: Conflict Management in a Divided World*, edited by C. A. Crocker, F. Osler Hampson and P. R. Aall. 53–65. Washington, DC: United States Institute of Peace Press, 2007.

Killmister, S. 'Remote Weaponry: The Ethical Implications'. *Journal of Applied Philosophy* 25, no. 2 (2008): 121–33.

Klosko, G. 'Contemporary Anglo-American Political Philosophy'. In *The Oxford Handbook of the History of Political Philosophy*, edited by G. Klosko. 456–79. Oxford: Oxford University Press, 2011.

Knightley, P. *The First Casualty: The War Correspondent as Hero and Myth-Maker from the Crimea to Iraq*. Baltimore: Johns Hopkins University Press, 2004.

Krahmann, E. *State, Citizens and the Privatization of Security.* Cambridge: Cambridge University Press, 2010.

Kreps, S. 'Flying under the Radar: A Study of Public Attitudes Towards Unmanned Aerial Vehicles'. *Research & Politics* 1, no. 1 (2014): 1–7.

Kreps, S. and J. Kaag. 'The Use of Unmanned Aerial Vehicles in Contemporary Conflict: A Legal and Ethical Analysis'. *Polity* 44, no. 2 (2012): 260–85.

Krishnan, A. 'Automating War: The Need for Regulation'. *Contemporary Security Policy* 30, no. 1 (2009): 172–93.

Krishnan, A. *Killer Robots: Legality and Ethicality of Autonomous Weapons*. Farnham: Ashgate, 2009.

Krishnan, A. *War as Business: Technological Change and Military Service Contracting.* Farnham: Ashgate, 2008.

Kurzweil, R. *The Singularity Is Near: When Humans Transcend Biology*. New York: Viking Penguin, 2005.

Kuschel, J. *The European Defense Market: Great Challenges – Small Moves*. Norderstedt: GRIN Verlag, 2008.

Lackey, D. P. *Moral Principles and Nuclear Weapons*. New Jersey: Rowman & Allanheld Publishers, 1984.

Lam, T. L. and Y. Xu. *Tree Climbing Robot: Design, Kinematics and Motion Planning*. Dordrecht: Springer, 2012.

Lamberth, I. 'K-Max – Unmanned Aerial Alternative to Convoys'. http://www.isaf.nato.int/article/news/k-max-unmanned-aerial-alternative-to-convoys.html.

Lendman, S. 'America's Drone Command Centers: Remote Warriors Operate Computer Keyboards and Joysticks'. Global Research: Centre for Research on Globalisation, http://www.globalresearch.ca/america-s-drone-command-centers-remote-warriors-operate-computer-keyboards-and-joysticks/30590.

Lepage, J. -D. *German Military Vehicles of World War II: An Illustrated Guide to Cars, Trucks, Half-Tracks, Motorcycles, Amphibious Vehicles and Others*. Jefferson: McFarland & Company, 2007.

Lin, P. 'Ethical Blowback from Emerging Technologies'. *Journal of Military Ethics* 9, no. 4 (2010): 313–31.

Lin, P. 'Introduction to Robot Ethics'. In *Robot Ethics: The Ethical and Social Implications of Robotics*, edited by P. Lin, K. Abney and G. Bekey. 3–16. Massachusetts: MIT Press, 2012.

Lin, P., K. Abney, and G. Bekey. 'Robot Ethics: Mapping the Issues for a Mechanized World'. *Artificial Intelligence* 175, no. 5–6 (2011): 942–49.

Lin, P., G. Bekey, and K. Abney. *Autonomous Military Robotics: Risk, Ethics, and Design*. San Luis Obispo: California Polytechnic State University, 2008.

Locke, J. *The Two Treatises of Government*. Indianapolis: Hackett Publishing Company, 1980.

Lokhorst, G. -J. and J. van den Hoven. 'Responsibility for Military Robots'. In *Robot Ethics: The Ethical and Social Implications of Robotics*, edited by P. Lin, K. Abney and G. Bekey. 145–56. Cambridge: MIT Press, 2012.

Loo, B. 'Revolutions in Military Affairs: Theory and Applicability to Small Armed Forces'. In *Military Transformation and Strategy: Revolutions in Military Affairs and Small States*, edited by B. Loo. 1–12. London: Taylor & Francis, 2008.

Lucas, G. 'Industrial Challenges of Military Robotics'. In *The Ethics of Emerging Military Technologies*. University of San Diego: International Society for Military Ethics, 2011.

Lucas, G. 'Industrial Challenges of Military Robotics'. *Journal of Military Ethics* 10, no. 4 (2011): 274–95.

Lucas, G. 'Postmodern War'. *Journal of Military Ethics* 9, no. 4 (2010): 289–98.

Lucas, G. 'The Ethical Challenges of Unmanned Systems'. In *Robots on the Battlefield: Contemporary Issues and Implications for the Future*, edited by R. Doare, D.

Danet and J. P. Hanon. 135–37. Fort Leavenworth: Combat Studies Institute Press, 2014.

Mack, A. 'Why Big Nations Lose Small Wars: The Politics of Asymmetric Conflict'. *World Politics: A Quarterly Journal of International Relations* 27, no. 2 (1975): 175–200.

Mahabharata, The. Translated by P. Chandra Roy. Calcutta: Bharata Press.

Mahnken, T. G. *Technology and the American Way of War.* New York: Columbia University Press, 2008.

Manchester, W. *Goodbye Darkness: A Memoir of the Pacific War.* New York: Dell Publishing Company, 1979.

Manders-Huits, N. 'Moral Responsibility and IT for Human Enhancement'. Paper presented at the Association for Computing Machinery Symposium on Applied Computing, Dijon, 2006.

Maqbool, A. 'Mapping US Drone and Islamic Militant Attacks in Pakistan'. *BBC News South Asia* (2010). Published electronically 22 July. http://www.bbc.co.uk/news/world-south-asia-10728844.

Margalit, A. and M. Walzer. 'Israel: Civilians & Combatants'. *The New York Review of Books* 56, no. 8 (2009). http://www.nybooks.com/articles/archives/2009/may/14/israel-civilians-combatants/.

Marshall, S. L. A. *Men against Fire: The Problem of Battle Command.* University of Oklahoma Press: Norman, 2000.

Martin, G. and E. Steuter. *Pop Culture Goes to War: Enlisting and Resisting Militarism in the War on Terror.* Plymouth: Lexington Books, 2010.

Martin, M. J. and C. W. Sasser. *Predator: The Remote-Control Air War over Iraq and Afghanistan: A Pilot's Story.* Minneapolis: Zenith Press, 2010.

Marx, R. F. *The History of Underwater Exploration.* Mineola: Dover Publications, 1990.

Matthias, A. 'Is the Concept of an Ethical Governor Philosophically Sound?' In *Tilting Perspectives: Technologies on the Stand: Legal and Ethical Questions in Neuroscience and Robotics.* Netherlands: Tilburg University, 2011.

Matthias, A. 'The Responsibility Gap: Ascribing Responsibility for the Actions of Learning Automata'. *Ethics and Information Technology* 6 (2004): 175–83.

McKenzie, K. *The Revenge of the Melians: Asymmetric Threats and the Next QDR.* Edited by Institute for National Strategic Studies Washington, DC: National Defense University, 2000.

McMahan, J. *Killing in War.* Oxford: Oxford University Press, 2009.

McMahan, J. 'Preventative War and the Killing of the Innocent'. In *The Ethics of War: Shared Problems in Different Traditions*, edited by R. Sorabji and D. Rodin. 169–90. Aldershot: Ashgate, 2006.

McMahan, J. 'The Ethics of Killing in War'. *Ethics* 114, no. 1 (2004): 693–733.

McMahan, J. and R. McKim. 'The Just War and the Gulf War'. *Canadian Journal of Philosophy* 23, no. 4 (1993): 501–41.

Meadors, T. 'Virtual Jus in Bello: Teaching Just War with Video Games'. In *The Ethics of Emerging Military Technologies*. University of San Diego: International Society of Military Ethics, 2011.

Meagher, R. E. *Herakles Gone Mad: Rethinking Heroism in an Age of Endless War*. Northampton: Olive Branch Press, 2006.

Meilinger, P. S. '10 Propositions Regarding Air Power'. Washington, DC: Air Force History and Museums Program, 1995.

Metz, S. and D. V. Johnson. 'Asymmetry and U.S. Military Strategy: Definition, Background, and Strategic Concepts'. Washington, DC: United States Strategic Studies Institute, 2001.

Michael, M. G. and K. Michael. 'Towards a State of Uberveillance'. *IEEE Technology and Society* 29, no. 2 (2010): 9–16.

Millar, J. and I. Kerr. 'Delegation, Relinquishment and Responsibilty: The Prospect of Expert Robots'. In *We Robot*. Coral Gables: University of Miami School of Law, 2012.

Miller, S. 'Collective Responsibility and Information and Communication Technology'. In *Information Technology and Moral Philosophy*, edited by J. van den Hoven and J. Weckert. 226–50. Cambridge: Cambridge University Press, 2008.

Miller, S. 'Technology and War'. *Bulletin of the Atomic Scientists* 41, no. 11 (1985): 46–48.

Mohammed, A. bin. 'The Al-Qaida Papers – Drones'. http://hosted.ap.org/specials/interactives/_international/_pdfs/al-qaida-papers-drones.pdf.

Moore, G. E. 'Cramming More Components onto Integrated Circuits'. *Electronics* 38, no. 8 (1965): 114–17.

Moore, S. 'The Federal Budget 10 Years Later: The Triumph of Big Government'. In *The Republican Revolution 10 Years Later: Small Government or Business as Usual?*, edited by C. R. Edwards and J. C. Samples. 59–72. Washington, DC: CATO Institute, 2005.

Moran, M. E. 'The History of Robotic Surgery'. In *Robotics in Genito-Urinary Surgery*, edited by A. K. Hemal and M. Menon. Dordrecht: Springer, 2010.

Moretz, J. *The Royal Navy and the Capital Ship in the Interwar Period: An Operational Perspective*. London: Frank Cass Publishers, 2002.

Morin, J. W. *Design, Fabrication and Mechanical Optimization of Multi-Scale Anisotrophic Feet for Terrestrial Locomotion*. Massachusetts: Massachusetts Institute of Technology, 2011.

Moynihan, R. 'Selling Sickness: The Pharmaceutical Industry and Disease Mongering'. *British Medical Journal* 324, no. 7242 (2002): 886–91.

Müller, H. 'No Need to Fear United Germany'. *Bulletin of the Atomic Scientists* 46, no. 3 (1990): 14–15.

Müller, H., M. Fey, S. Mannitz, and N. Schörnig. *Demokratie, Streitkräfte Und Militärische Einsätze: Der Zweite Gesellschaftsvertrag Steht Auf Dem Spiel.* Frankfurt: Hessische Stiftung Friedens-und Konfliktforschung, 2010.

Murnion, W. E. 'A Postmodern View of Just War'. In *Intervention, Terrorism, and Torture: Contemporary Challenges to Just War Theory*, edited by S. P. Lee. Dordrecht: Springer, 2007.

Murphie, A. and J. Potts. *Culture and Technology*. London: Palgrave, 2003.

Murphy, J. 'The Killing of the Innocent'. In *War, Morality, and the Military Profession*, edited by M. M. Wakin. Boulder, CO: Westview, 1986.

Murray, W. 'Thinking About Revolutions in Military Affairs'. *Joint Force Quarterly* 16 (1997): 69–76.

Nagel, T. 'War and Massacre'. In *War, Morality, and the Military Profession*, edited by M. M. Wakin. 297–316. Boulder, CO: Westview Press, 1986.

Nathanson, S. *Terrorism and the Ethics of War.* Cambridge: Cambridge University Press, 2010.

National Research Council (US). Committee on the Effects of Nuclear Earth-Penetrator and Other Weapons. *Effects of Nulcear Earth-Penetrator and Other Weapons.* Washington, DC: National Academies Press, 2005.

Needham, J. *Science and Civilisation in China: Chemistry and Chemical Technology.* Cambridge: Cambridge University Press, 1987.

Needham, J. *Science and Civilisation in China: Physics and Physical Technology.* Cambridge: Cambridge University Press, 1965.

Needham, J. *Science and Civilisation in China: History of Scientific Thought.* Translated by L. Wang. Cambridge: Cambridge University Press, 1956.

Neff, S. C. *War and the Law of Nations: A General History.* Cambridge: Cambridge University Press, 2005.

Nemeth, M. 'Zyvex Technologies Reveals the Piranha Unmanned Surface Vessel'. http://www.prweb.com/releases/zyvex/piranhaUSV/prweb4658434.htm.

Newcome, L. R. *Unmanned Aviation: A Brief History of Unmanned Aerial Vehicles.* Reston: American Institute of Aeronautics and Astronautics, 2004.

Nonami, K., F. Kendoul, S. Suzuki, et al. *Autonomous Flying Robots.* Dordrecht: Springer, 2010.

Noorman, M. 'Computing and Moral Responsibility'. http://plato.stanford.edu/archives/fall2012/entries/computing-responsibility/.

Norman, R. *Killing, Ethics, and War.* Cambridge: Cambridge University Press, 1995.

Nygren, K. P. 'Emerging Technologies and Exponential Change: Implications for Army Transformation'. *Parameters: The US Army War College Quarterly Senior Professional Journal* 32, no. 2 (2002): 86–99.

O'Brien, W. V. *The Conduct of a Just and Limited War.* New York: Praeger, 1981.

O'Brien, W. V. 'The Conduct of Just and Limited War'. In *Contemporary Moral Problems*, edited by J. E. White. Albany: Thomson-Wadsworth, 2009.

O'Brien, W. V. and J. Langan, eds. *The Nuclear Dilemma and the Just War Tradition*. Lexington, MA: Lexington Books, 1986.

O'Connell, C. 'Drone Command Center, 200-Plus Jobs Coming to Horsham'. Fox News, http://www.myfoxphilly.com/story/21676061/drone-command-center-200-jobs-coming-to-horsham.

O'Hanlon, M. *Technological Change and the Future of Warfare*. Washington, DC: Brookings Institution Press, 2000.

O'Hanlon, M. and P. W. Singer. 'The Real Defense Budget Questions'. *Politico*, 21 July 2011, 1–2.

O'Keefe, M. and C. A. J. Coady, eds. *Terrorism and Justice: Moral Argument in a Threatened World*. Carlton: Melbourne University Press, 2002.

Ofek, H. 'The Tortured Logic of Obama's Drone War'. *The New Atlantis: A Journal of Technology and Society* 27, no. 1 (2010): 35–44.

Office of the Press Secretary. 'Remarks by the President in Address to the Nation on the Way Forward in Afghanistan and Pakistan'. The White House, http://www.whitehouse.gov/the-press-office/remarks-president-address-nation-way-forward-afghanistan-and-pakistan.

Office of the Surgeon General. 'Mental Health Advisory Team (MHAT) IV Operation Iraqi Freedom 05–07, Final Report'. Washington, DC: United States Department of the Army, 2006.

Office of the Under Secretary of Defense (Comptroller). *Program Acquisition Cost by Weapon System*. Washington, DC: Department of Defense, 2013.

Orend, B. 'Jus Post Bellum: A Just War Theory Perspective'. In *Just Post Bellum: Towards a Law of Transition from Conflict to Peace*, edited by S. Carsten and J. K. Kleffner. 31–52. The Hague: TMC Aasser Press, 2008.

Orend, B. *The Morality of War*. Peterborough: Boradview Press, 2006.

Orend, B. *War and International Justice: A Kantian Perspective*. Waterloo: Wilfrid Laurier University Press, 2000.

Osofsky, M. J., A. Bandura, and P. G. Zimbardo. 'The Role of Moral Disengagement in the Execution Process'. *Law and Human Behavior* 29, no. 4 (2005): 371–93.

Owens, P. 'Distinctions, Distinctions: 'Public' and 'Private' Force?' *International Affairs* 84, no. 5 (2008): 977–90.

Patterson, E. D. *Ending Wars Well: Order, Justice, and Conciliation in Contemporary Post-Confict*. New Haven: Yale University Press, 2012.

Payne, K. 'The Media as an Instrument of War'. *Parameters: The US Army War College Quarterly Senior Professional Journal* 35, no. 1 (2005): 81–93.

Perkowitz, S. *Digital People: From Bionic Humans to Androids*. Washington, DC: Joseph Henry Press, 2005.

Peters, R. *Fighting for the Future: Will American Triumph?* Pennsylvania: Stackpole Books, 2001.

Petitt, P. 'Responsibility Incorporated'. *Ethics* 117, no. 2 (2007): 171–201.

Pew Research. 'Drone Strikes Widely Opposed'. http://www.pewglobal.org/2012/06/13/global-opinion-of-obama-slips-international-policies-faulted/.

Phillips, R. L. *War and Justice*. Norman: University of Oklahoma Press, 1984.

Pierce, A. C. 'War, Strategy, and Ethics'. In *Ethics and the Future of Conflict: Lessons from the 1990s*, edited by A. Lang, A. C. Pierce and J. H. Rosenthal. 9–17. Upper Saddle River: Prentice Hall, 2004.

Plaw, A. 'Counting the Dead: The Proportionality of Predation in Pakistan'. In *Killing by Remote Control: The Ethics of an Unmanned Military*, edited by B. J. Strawser. 126–53. New York: Oxford, 2013.

Plaw, A., M. Fricker, and B. G. Williams. 'Practice Makes Perfect?: The Changing Civilian Toll of CIA Drone Strikes in Pakistan'. *Perspectives on Terrorism* 5, no. 5–6 (2011): 51–69.

Plaw, A., I. Mothana, J. Cavallaro, et al. 'Do Drone Attacks Do More Harm Than Good'. *The New York Times* (2012). http://www.nytimes.com/roomfordebate/2012/09/25/do-drone-attacks-do-more-harm-than-good.

Pseudo-Apollodorus. *The Library of Greek Mythology*. Translated by K. Aldrich. Lawrence: Coronado Press, 1975.

Pseudo-Hyginus. *The Myths of Hyginus*. Translated by M. A. Grant. Lawrence: University of Kansas Press, 1960.

Pufendorf, S. *On the Law of Nature and Nations*. Translated by C.H. Oldfather. Oxford: Clarendon Press, 1934.

Pushies, F. J. *Night Stalkers: 160th Special Operations Aviation Regiment (Airborne)*. Minnesota: Zenith, 2005.

Pushies, F. J. *U.S. Air Force Special Ops*. Minneapolis: Zenith Imprint, 2007.

Qiao, L. and X. Wang. *Unrestricted Warfare: China's Master Plan to Destroy American*. West Palm Beach: NewsMax Media, 2002.

Quinn, B. 'Mod Study Sets out How to Sell Wars to the Public'. *The Guardian* (2013). http://www.theguardian.com/uk-news/2013/sep/26/mod-study-sell-wars-public.

Raghavan, S. 'In Yemen, Us Airstrikes Breed Anger, and Sympathy for Al Qaeda'. http://articles.washingtonpost.com/2012-05-29/world/35456187_1_aqap-drone-strikes-qaeda.

Raibert, M., K. Blankespoor, G. Nelson, and R. Playter. 'Bigdog, the Rough-Terrain Quadruped Robot'. Paper presented at the 17th World Congress, Seoul, Korea, 6–11 July 2008.

Ramsey, P. *The Just War: Force and Political Responsibility*. New York: Charles Scribner's Sons, 1968.

Raugh, H. E. *The Victorians at War, 1815–1914: An Encyclopedia of British Military History*. Santa Barbara: ABC-CLIO, 2004.

Rawls, J. *A Theory of Justice*. New York: Oxford University Press, 1999.

Rawls, J. *Political Liberalism*. New York: Columbia University Press, 2005.

Rawls, J. *The Law of Peoples*. Cambridge: Harvard University Press, 2001.

Reidy, C. 'Navy Doubles Foster-Miller's Robot Contract'. *Boston Globe*, 2007.

Replogie, J. 'The Drone Makers and Their Friends in Washington'. http://www.fronterasdesk.org/news/2012/jul/05/drone-makers-friends-washington/#.UBfn8FEWH0c.

Robben, A. C. G. M. 'Losing Hearts and Minds in the War on Terrorism'. In *Iraq at a Distance: What Anthropologists Can Teach Us About the War*, edited by A. C. G. M. Robben. 106–32. Philadelphia: University of Pennsylvania Press, 2010.

Roderick, I. 'Considering the Fetish Value of EOD Robots: How Robots Save Lives and Sell War'. *International Journal of Cultural Studies* 13, no. 3 (2010): 235–53.

Rodin, D. 'The Ethics of Asymmetric War'. In *The Ethics of War: Shared Problems in Different Traditions*, edited by R. Sorabji and D. Rodin. 153–68. Aldershot: Ashgate, 2006.

Rodin, D. *War and Self-Defense*. Oxford: Clarendon, 2002.

Rodriguez, A. 'Pakistani Death Squads Go after Informants to US Drone Program'. http://articles.latimes.com/2011/dec/28/world/la-fg-pakistan-death-squads-20111228.

Rogers, W. and S. Rogers. *Storm Center: The USS Vincennes and Iran Air Flight 655*. Annapolis: Naval Institute Press, 1992.

Rosheim, M. E. *Leonardo's Lost Robots*. Dordrecht: Springer, 2006.

Rousseau, J. -J. *On the Social Contract*. Translated by G. D. H. Cole. Mineola: Courier Dover Publications, 2003.

Royakkers, L. and R. van Est. 'The Cubicle Warrior: The Marionette of Digitalized Warfare'. *Ethics and Information Technology* 12, no. 3 (2010): 289–96.

Royal, R. 'In My Beginning Is My End'. In *Ethics Beyond War's End*, edited by E. D. Patterson. 65–76. Washington, DC: Georgetown, 2012.

Rush, R. S. *Enlisted Soldier's Guide*. Mechanicsburg: Stackpole Books, 2006.

Safire, W. *The Right Word in the Right Place at the Right Time: Wit and Wisdom from the Popular 'on Language' Column in the New York Times Magazine*. New York: Simon & Schuster, 2004.

Samuel, A. and J. Weir. *Introduction to Engineering Design: Modelling, Synthesis and Problem Solving Strategies*. Burlington: Elsevier Butterworth-Heinemann, 2005.

Searle, J. 'Minds, Brains, and Programs'. *The Behavioral and Brain Sciences* 3, no. 3 (1980): 417–57.

Sellers, D. P., J. A. Ramsbotham, H. Bertrand, and N. Karvonides. *International Assessment of Unmanned Ground Vehicles*. Alexandria: Institute for Defense Analyses, 2008.

Shactman, N. 'Robot Canon Kills 9, Wounds 14'. http://www.wired.com/dangerroom/2007/10/robot-cannon-ki/.

Shactman, N. 'Robot Sick Bay'. *Popular Mechanics* 183, no. 7 (2006): 18–19.

Shaker, S. M. and A. R. Wise. *War without Men: Robots on the Future Battlefield.* Washington: Pergamon-Brassey's, 1988.

Sharkey, N. 'I Ropebot'. *New Scientist* 194, no. 2611 (2007): 32–35.

Sharkey, N. 'Moral and Legal Aspects of Military Robots'. In *Ethical and Legal Aspects of Unmanned Systems: Interviews*, edited by G. Dabringer. 43–51. Vienna: Institute for Religion and Peace, 2010.

Sharkey, N. 'Saying 'No!' To Lethal Autonomous Targeting'. *Journal of Military Ethics* 9, no. 4 (2010): 369–83.

Shaw, M. *The New Western Way of War: Risk-Transfer War and Its Crisis in Iraq.* Cambridge: Polity Press, 2005.

Shue, H. 'Do We Need a 'Morality of War'?' In *Just and Unjust Warriors: The Moral and Legal Status of Soldiers*, edited by D. Rodin and H. Shue. 87–111. Oxford: Oxford University Press, 2008.

Shue, H. 'Mediating Duties'. *Ethics* 98, no. 4 (1988): 687–704.

Shurtleff, K. 'The Effects of Technology on Our Humanity'. *Parameters: The US Army War College Quarterly Senior Professional Journal* 32, no. 2 (2002): 100–12.

Siciliano, B. and O. Khatib, eds. *The Springer Handbook of Robotics.* Dordrecht: Springer, 2008.

Simpson, R. and R. Saparrow. 'Nanotechnologically Enhanced Combat Systems: The Downside of Invulnerability'. In *In Pursuit of Nanoethics*, edited by B. Gordijn and A. Cutter. 89–103. Dordrecht: Springer, 2013/14.

Singer, P. W. 'The Ethics of Killer Applications: Why Is It So Hard to Talk About Morality When It Comes to New Military Technology?' *Journal of Military Ethics* 9, no. 4 (2010): 299–312.

Singer, P. W. 'The Future of War'. In *Ethical and Legal Aspects of Unmanned Systems: Interviews*, edited by G. Dabringer. 71–84. Vienna: Institute for Religion and Peace, 2010.

Singer, P. W. 'Military Robots and the Laws of War'. *The New Atlantis: A Journal of Technology and Society* 23 (2009): 27–47.

Singer, P. W. 'U-Turn: Unmanned Systems Could Be Casualties of Budget Pressures'. *Armed Forces Journal* (April 2011).

Singer, P. W. *Wired for War: The Robotics Revolution and Conflict in the 21st Century.* New York: The Penguin Press, 2009.

Sjoberg, L. *Gender, Justice, and the Wars in Iraq: A Feminist Reformulation of Just War Theroy.* Lanham: Lexington Books, 2006.

Sjoberg, L. 'Why Just War Needs Feminism Now More Than Ever'. *International Politics* 45, no. 1 (2008): 1–18.

Slipchenko, V. and A. Khokhlov. 'Shock and Awe: Russian Expert Predicts 500,000 Iraqi Dead in War Designed to Test Weapons'. http://globalresearch.ca/articles/SLI303A.html.

Sloan, E. C. *The Revolution in Military Affairs: Implications for Canada and Nato.* McGill-Queen's University Press, 2002.

Sluka, J. A. 'Drones in the Tribal Zone: Virtual and Losing Hearts and Minds in the Af-Pak War'. In *War, Technology, Anthropology*, edited by K. Stroeken. 21–33. New York: Berghahn Books, 2012.

Smit, W., ed. *Just War and Terrorism: The End of the Just War Concept?* Leuven: Peeters Publishers, 2005.

Snider, D., J. Nagl, and T. Pfaff. *Army Professionalism, the Military Ethic, and Officership in the 21st Century*. Carlisle: The Strategic Studies Institute, 1999.

Solomon, N. 'The Ethics of War: Judaism'. In *The Ethics of War: Shared Problems in Different Traditions*, edited by R. Sorabji and D. Rodin. 108–37. Aldershot: Ashgate, 2006.

Sparrow, R. 'Building a Better Warbot: Ethical Issues in the Design of Unmanned Systems for Military Applications'. *Science and Engineering Ethics* 15, no. 2 (June 2009): 169–87.

Sparrow, R. 'Can Machines Be People: Relfections on the Turing Triage Test'. In *Robot Ethics: The Ethical and Social Implications of Robotics*, edited by Pa. Lin, K. Abney and G. Bekey. 301–16. Cambridge: MIT Press, 2012.

Sparrow, R. 'Killer Robots'. *Journal of Applied Philosophy* 24, no. 1 (2007): 62–77.

Sparrow, R. 'Predators or Plowshares? Arms Control of Robotic Weapons'. *IEEE Technology and Society* 28, no. 1 (2009): 25–29.

Sparrow, R. 'Robotic Weapons and the Future of War'. In *New Wars and New Soldiers*, edited by J. Wolfendale and P. Tripodi. 117–33. Farnham: Ashgate, 2011.

Sparrow, R. 'The Ethical Challenges of Military Robots'. In *Ethical and Legal Aspects of Unmanned Systems: Interviews*, edited by G. Dabringer. 87–101. Vienna: Institute for Religion and Peace, 2010.

Springer, P. *Military Robots and Drones*. Santa Barbara: ABC-CLIO, 2013.

Stahn, C. 'Jus Post Bellum: Mapping the Discipline(s)'. In *Just Post Bellum: Towards a Law of Transition from Conflict to Peace*, edited by C. Stahn and J. K. Kleffner. 93–114. The Hague: TMC Aasser Press, 2008.

Stiglitz, J. and L. Bilmes. *The Three Trillion Dollar War*. New York: W. W. Norton & Company, 2008.

Stine, D. D. *Federally Funded Innovation Inducement Prizes*. Darby: DIANE Publishing, 2009.

Story, E. 'Complacency = Mishap'. *Combat Edge* 15, no. 4 (2006): 10–11.

Strangelove, M. *Watching YouTube: Extraordinary Videos by Ordinary People*. Ontario: University of Toronto Press, 2010.

Strawser, B. J. 'Introduction: The Moral Landscape of Unmanned Weapons'. In *Killing by Remote Control: The Ethics of an Unmanned Military*, edited by B. J. Strawser. 3–24. New York: Oxford University Press, 2013.

Strawser, B. J., ed. *Killing by Remote Control: The Ethics of an Unmanned Military*. New York: Oxford University Press, 2013.

Strawser, B. J. 'Moral Predators: The Duty to Employ Uninhabited Aerial Vehicles'. *Journal of Military Ethics* 9, no. 4 (2010): 342–68.

Strawser, B. J. 'Two Bad Arguments for the Justification of Autonomous Weapons'. In *Technology and Security*. University of North Texas: Society for Philisophy and Technology, 2011.

Strawson, P. F. 'Freedom and Resentment'. In *Freedom and Resentment and Other Essays*. Methuen: London, 1974.

Stuckart, D. W. and M. J. Berson. 'Artificial Intelligence in the Social Studies'. In *Research on Technology in Social Studies Education*, edited by J. Lee and A. M. Friedman. Charlotte: Information Age Publishing, 2009.

Suárez, F. 'On War'. In *Francisco Suárez, Selections from Three Works*, edited by G. L. Williams, Ammi Brown and John Waldron. Oxford: Clarendon Press, 1944.

Suárez, F. 'The Three Theological Virtues'. In *The Classics of International Law*, edited by J. Brown Scott Washington, DC: Institute of International Law, 1944.

Sullins, J. P. 'Aspects of Telerobotic Systems'. In *Ethical and Legal Aspects of Unmanned Systems: Interviews*, edited by G. Dabringer. 157–67. Vienna: Institute for Religion and Peace, 2010.

Sullins, J. P. 'Introduction: Open Questions in Roboethics'. *Journal of Philosophy and Technology* 24, no. 3 (2011): 233–38.

Sullins, J. P. 'Robowarfare: Can Robots Be More Ethical Than Humans on the Battlefield?' *Ethics and Information Technology* 12, no. 3 (2010): 263–75.

Sullins, J. P. 'When Is a Robot a Moral Agent?' *International Review of Information Ethics* 6, no. 1 (2006): 24–30.

Sullivan, J. M. 'Defense Acquisitions: DOD Could Achieve Greater Commonality and Efficiencies among Its Unmanned Aircraft Systems'. edited by Subcommittee on National Security and Foreign Affairs and Committee on Oversight and Government Reform. Washington, DC: Government Accountability Office, 2010.

Taylor, G., ed. *Henry V*. Oxford: Oxford University Press, 1982.

Temes, P. S. *The Just War*. Chicago: Ivan R. Dee, 2003.

Tenner, E. *Why Things Bite Back: Technology and the Revenge of Unintended Consequences*. New York: Vintage Books, 1996.

The United States Joint Staff. *Joint Strategy Review*. Washington, DC: Department of Defense, 1999.

Thompson, D. 'Moral Responsibility and Public Officials: The Problem of Many Hands'. *American Political Science Review* 74, no. 4 (1980): 905–16.

Thompson, D. *Political Ethics and Public Office*. Cambridge: Harvard University Press, 1987.

Thompson, L. and D. J. Gillan. 'Social Factors in Human-Robot Interaction'. In *Human-Robot Interactions in Future Military Operations*, edited by M. Barnes and F. Jentsch. 67–82. Farnham: Ashgate, 2011.

Treverton, G. F. *Reshaping National Intelligence for an Age of Information*. Cambridge: Cambridge University Press, 2003.
Turse, N. *The Complex: How the Military Invades Our Everyday Lives*. London: Faber & Faber, 2008.
Tyner, J. A. *Military Legacies: A World Made by War*. New York: Routledge, 2010.
Tzu, S. *The Art of War*. London: Wordsworth Editions, 1990.
United Kingdom Ministry of Defence. 'Joint Doctrine Note 2/11: The UK Approach to Unmanned Aircraft Systems'. 1–102. Wiltshire: Ministry of Defence, 2011.
United States Army. 'Field Manual'. Washington, DC: Department of Defense, 1994.
United States Catholic Bishops. 'The Just War and Non-Violent Positions'. In *War, Morality, and the Military Profession*, edited by M. Wakin. Boulder, CO: Westview Press, 1989.
United States Department of Defense. 'FY2009–2034 Unmanned Systems Integrated Roadmap'. 1–210. Washington, DC: Department of Defense, 2009.
United States Department of Defense. 'Unmanned Systems Integrated Roadmap FY2011–2036'. 1–89. Washington, DC: Department of Defense, 2011.
United States Department of Defense. 'Unmanned Systems Integrated Roadmap FY2013–2038'. 1–89. Washington, DC: Department of Defense, 2013.
United States Environmental Protection Agency. 'EPA's Budget and Spending'. http://www.epa.gov/planandbudget/budget.html.
United States National Research Council. *Star 21: Strategic Technologies for the Army of the Twenty-First Century*. Washington, DC: National Academy Press, 1993.
United States National Research Council, Committee on Autonomous Vehicles in Support of Naval Operations. *Autonomous Vehicles in Support of Naval Operations*. Washington, DC: The National Academies Press, 2005.
Unknown. 'Who Called Them "New"?' *Popular Science*, June 1945, 1.
Van Damme, G. and N. Fotion. 'Proportionality'. In *Moral Constraints on War: Principles and Cases*, edited by B. Coppieters and N. Fotion. 159–70. Lanham Lexington, 2008.
van den Berg, R. 'The 21st Century Battlespace: The Danger of Technological Ethnocentrism'. *Canadian Military Journal* 10, no. 4 (2010): 10–18.
van Wifferen, L. 'Alienation from the Battlefield: Ethical Considerations Concerning Remote Controlled Military Robotics'. Universiteit Utrecht, 2011.
van Wynsberghe, A. *Care Robots: An Ethical Framework*. Farnham: Ashgate (forthcoming).
Vanya, L. 'Excepts from the History of Unmanned Ground Vehicles'. *AARMS* 2, no. 2 (2003): 185–97.
Vassallo, S. V. and L. S. Nelon. 'Thermoregulatory Principles'. In *Goldfrank's Toxicologic Emergencies*, edited by M. J. Wonsiewicz, K. G. Edmonson and P. J. Boyle. 255–76. New York: McGraw Hill, 2006.

Veruggio, G. and F. Operto. 'Roboethics: Social and Ethical Implications of Robotics'. In *Springer Handbook of Robotics*, edited by B. Siciliano and O. Khatib. 1499–524. Dordrecht: Springer Verlag, 2008.

Vickers, M. G. and R. C. Martinage. *The Revolution in War*. Washington, DC: Center for Strategic and Budgetary Assessments, 2004.

Victoria, F. de. 'De Indis Et De Ivre Belli Relectiones'. In *The Classics of International Law*, edited by J. Brown Scott and E. Nys. Washington, DC: Institute of International Law, 1917.

Vieira, W. G. R., C. De Marqui, A. Erturk, and D.J. Inman. 'Structural Dynamics and Renewable Energy'. Paper presented at the 28th IMAC, A Conference on Structural Dynamics, 2010.

von Clausewitz, C. *On War*. Translated by O. J. Matthijis Jolles. New York: The Modern Library, 1943.

von Clausewitz, C. *Principles of War*. Translated by H. W. Gatzke. Harrisburg: The Military Service Publishing Company, 1942.

Waelbers, K. 'Technological Delegation: Responsibility for the Unintended'. *Science and Engineering Ethics* 15, no. 1 (March 2009): 51–68.

Wagner, M. 'The Dehumanization of International Humanitarian Law: Independently Operating Weapon System and Modern Armed Conflict'. In *We Robot*. Coral Gables: University of Miami School of Law, 2012.

Walker, P. D. *Battle Fatigue: Understanding PTSD and Finding a Cure*. Bloomington: iUniverse, 2008.

Wall, T. and T. Monahan. 'Surveillance and Violence from Afar: The Politics of Drones and Liminal Security-Scapes'. *Theoretical Criminology* 15, no. 3 (2011): 239–54.

Wallach, W. and C. Allen. *Moral Machines: Teaching Robots Right from Wrong*. Oxford: Oxford University Press, 2009.

Walzer, M. *Just and Unjust Wars: A Moral Argument with Historical Illustrations*. New York: Basic Books, 2006.

Walzer, M. 'Response to McMahan's Paper'. *Philosophia* 34, no. 1 (2006): 43–45.

Walzer, M. 'The Aftermath of War'. In *Ethics Beyond War's End*, edited by E. D. Patterson. 35–46. Washington, DC: Georgetown University Press, 2012.

Watson, P. *War on the Mind: The Military Uses and Abuses of Psychology*. New York: Basic Books, 1978.

Weckert, J. 'Risks and Scientific Responsibilities in Nanotechnology'. In *Handbook of Risk Theory: Epistemology, Decision Theory, Ethics, and Social Implications of Risk*, edited by S. Roeser, R. Hillberbrand, P. Sandin and M. Peterson. 160–75. Dordrecht: Springer, 2011.

Weiner, T. 'GI Robot Rolls toward the Battlefield'. *Herald Tribune*, 17 February 2005.

Weizenbaum, J. *Computer Power and Human Reason: From Judgement to Calculation*. New York: W. H. Freeman & Company, 1977.

Werrell, K. P. 'Did USAF Technology Fail in Vietnam?' *Airpower Journal* 12, no. 1 (1998): 87–99.

Wertheim, E. *Naval Institute Guide to Combat Fleets of the World: Their Ships, Aircraft, and Systems*. Annapolis: Naval Institute Press, 2007.

Whetham, D. 'Remote Killing and Drive-by Wars'. In *Protecting Civilians During Violent Conflict: Theoretical and Practical Issues for the 21st Century*, edited by David W. Lovell and I. Primoratz. 199–214. Farnham: Ashgate, 2012.

Wickersham, E. E. 'Land Torpedo'. United States Patent, 1922.

Williams, H. *Kant and the End of War: A Critique of Just War Theory*. New York: Palgrave Macmillan, 2012.

Williamson, G. *German E-Boats, 1939–45*. University Park: Osprey Publishing, 2002.

Wilson, S., G. Miller, and S. Horwitz. 'Boston Bombing Suspect Cites U.S. Wars as Motivation, Officials Say'. *The Washington Post* (2013). http://articles.washingtonpost.com/2013-04-23/national/38751370_1_u-s-embassy-boston-marathon-bombings.

Xie, M. *The Fundamentals of Robotics: Linking Perception to Action*. River Edge: World Scientific, 2003.

Yenne, B. *Attack of the Drones: A History of Unmanned Aerial Combat*. Minneapolis: Zenith Press, 2004.

YouTube. 'Hell Is Coming for Breakfast'. http://www.youtube.com/watch?v=bHCchnGdtJA.

YouTube. 'UAV Kills 6 Heavily Armed Criminals'. http://www.youtube.com/watch?v=gNNJJrcIa7A&list=PL5FC6E7FB6B2FA591&index=4&feature=plpp_video.

Zaloga, S. J. *Unmanned Aerial Vehicles: Robotic Air Warfare 1917–2007*. Oxford: Osprey Publishing, 2008.

Zaloga, S. J. *V-1 Flying Bomb 1942–52: Hitler's Infamous 'Doodlebug'*. University Park: Osprey Publishing, 2005.

Zenko, M. '10 Things You Didn't Know About Drones'. *Foreign Policy* March/April (2012). http://www.foreignpolicy.com/articles/2012/02/27/10_things_you_didnt_know_about_drones?page=full.

Zuboff, S. 'Automate/Informate: The Two Faces of Intelligent Technology'. *Organizational Dynamics* 14, no. 2 (1985): 5–18.

Zupan, D. S. 'A Presumption of the Moral Equality of Combatants: A Citizen-Soldier's Perspective'. In *Just and Unjust Warriors: The Moral and Legal Status of Soldiers*, edited by D. Rodin and H. Shue. 214–25. Oxford: Oxford University Press, 2008.

Index

For Product Safety Concerns and Information please contact our EU
representative GPSR@taylorandfrancis.com
Taylor & Francis Verlag GmbH, Kaufingerstraße 24, 80331 München, Germany

www.ingramcontent.com/pod-product-compliance
Ingram Content Group UK Ltd.
Pitfield, Milton Keynes, MK11 3LW, UK
UKHW020013200926
475702UK00014B/1859

* 9 7 8 1 1 3 8 0 7 9 7 1 7 *